W9-CFO-297

The Telecommunications Information Millennium

A Vision and Plan for the Global Information Society

Other books by Robert K. Heldman

Telecommunications Management Planning

ISDN in the Information Marketplace

Future Telecommunications: Information Applications, Services, & Infrastructure

Global Telecommunications: Layered Networks' Layered Services

Information Telecommunications: Networks, Products, and Services

The Telecommunications Information Millennium

A Vision and Plan for the Global Information Society

Robert K. Heldman

with contributions by

Thomas A. Bystrzycki

McGraw-Hill, Inc.

New York San Francisco Washington, D.C. Auckland Bogotá
Caracas Lisbon London Madrid Mexico City Milan
Montreal New Delhi San Juan Singapore
Sydney Tokyo Toronto

HE
7631
H454
1995

Library of Congress Cataloging-in-Publication Data

Heldman, Robert K.
   The telecommunications information millennium : a vision & plan for
the global information society / by Robert K. Heldman.
      p. cm.
   Includes index.
   ISBN 0-07-028106-8
   1. Telecommunication—Social aspects. 2. Telecommunication-
-Technological innovations. 3. Information networks—Social
aspects. I. Title.
HE7631.H454 1995
384—dc20
                                                    95-8324
                                                       CIP

pbk 1 2 3 4 5 6 7 8 9 DOC/DOC 9 9 8 7 6 5

ISBN 0-07-028106-8

*The sponsoring editor of this book was Steve Chapman. The executive
editor was Robert E. Ostrander, the book editor was Sally Anne Glover,
and the production supervisor was Katherine G. Brown. This book was
set in ITC Century Light. It was composed in Blue Ridge Summit, Pa.*

*Printed and bound by R. R. Donnelley & Sons Company, Crawfordsville,
Indiana.*

                                                       MH95

To my loving wife,
Valerie.

Contents

Part 3
Plan of action

Foreword

Thomas A. Bystrzycki

As we approach the new millennium, it is important to consider the numerous opportunities offered by the exciting, wide spectrum of forthcoming telecommunications information services. They will impact every aspect of our society. In the past, we have mainly concentrated our efforts in establishing more efficient and effective voice dial-tone offerings. However, as we near the end of this century of rapidly expanding technology, we need to reassess our efforts and refocus them to take better advantage of these exciting technical breakthroughs. We need to formulate a vision of increased breadth and scope that challenges us to expand our capabilities and enables anyone to communicate to anyone—anyway, anywhere, anytime through a full range of voice, data, image, text, graphic, and video information.

With this in mind, I encourage you to take the time to pause and carefully review this analysis of the technical possibilities and market opportunities available from a new family of narrowband, wideband, and broadband networks, products, and services. As you will see, this is a crisp, clear, concise assessment and plan for achieving the infrastructure necessary to support the full range of these exciting new services, which will flow along the information highway to the information marketplace throughout the third millennium—the telecommunications information millennium. I wish you enjoyable reading.

Acknowledgments

This final analysis brings the telecommunication series to a close. Over the years, many of the leaders of the forthcoming communications and computer (C&C) networks, products, and services have shared with us their views and visions for the future. They have participated in trials of the potential new services, such as the COMPASS trial described in appendix A, which many believe was the technology trial of the century. All in all, it has been an exciting period in the search for the right leading-edge technology to support a better society with a better quality of life.

Hopefully, this series of works has established a platform for further thinking. I wish to thank the hundreds and hundreds of planners and researchers who have taken the time to share their thoughts and exchange ideas on what could be, would be, and should be obtainable. They have come from the full cross section of both the academic and the business worlds, as well as from the multitude of users. Together we have twisted this or that particular feature or service to best solve their particular application needs. Many of these colleagues are from AT&T, Bell Laboratories, the various RBOCs, Bellcore, GTE, Siemens, Fujitsu, DSC, Alcatel, North American Rockwell, Boeing, IBM, DEC, ITT, Apple, Motorola, Control Data, 3M, Mayo Clinic, and others too numerous to mention here.

I especially wish to thank the leaders of U S WEST with whom I have worked over the years, especially Tom Bystrzycki, without whose continued support this final analysis could not have been accomplished so successfully. And, of course, I thank my son, Peter, whose technical and marketing advice and counsel was invaluable in helping to address the proper issues with the proper focus and clarity.

Again, I wish to thank my current and past editors within McGraw-Hill for their continued assistance and support: Larry Hager, Neil Levine, Steve Chapman, and Sally Anne Glover, as well as the manuscript artists and assistants such as Lois Gower and Becky Strom, who have made all these endeavors achievable. I wish you, the players of the information game, many future successes and accomplishments in your unending pursuit of establishing the information society in the forthcoming telecommunications information millennium.

Introduction

"There is a tide in the affairs of men
Which taken at the flood leads on to fortune;
Omitted, all the voyage of their life
Is bound in shallows and in miseries.
On such a full sea are we now afloat,
And we must take the current when it serves,
Or lose our ventures."

William Shakespeare

This is really a book within a book. The initial section coalesces and summarizes many of the key insights contained in my earlier works concerning the potential narrowband-wideband-broadband information networks and the services they might offer. Upon this foundation is overlaid and interlaced a challenging vision and plan for successfully achieving these technical possibilities as both providers and suppliers attempt to pursue the exciting new opportunities of the forthcoming information marketplace.

There have been considerable advances in technology in the many years since the breakup of the heavily regulated telecommunications industry monopoly. The past has given way to a more complex, challenging, competitive arena, but this high-stakes "OK Corral" shoot-out form of global competition, fostered by media hype and Wall Street analysts, need not simply focus on supposedly quick, near-term financial paybacks. There is now, more than ever, a need for establishing the appropriate solid infrastructure for sustaining long-term growth; upon this infrastructure, numerous new services from thousands of information service providers (ISPs), enhanced service providers (ESPs), and source database servers (SDBSs) can grow and blossom. If constructed in an orderly, phased, layered approach with the proper funding, the potential of fully integrated voice, data, text, image, and video offerings can be realized, enabling much, much more than could ever hope to be achieved.

This analysis addresses the issues raised by the following conversation, once overheard at an information users' convention. The discussion went something like this:

> Yes, of course we are building the information highway that Washington has been championing for several years now. Sure, we need to interconnect the universities with our government's high-tech research facilities, using large information pipes. Sort of like a larger version of Internet. Internet? Who says Internet doesn't work? What do you mean? Haven't you ever used Internet? Everyone at school is now accessing its bulletin boards. It's great and cheap! The Department of Defense funded it for many years, and then left it for the private sector. All sorts of new services are coming on it! Just the other day another bank announced it was going to send money transactions on it. Yeah, all sorts of things are coming. It's part of the information highway to the information society.
>
> Why, we are even going to get access to all kinds of information from our new 500-channel cable systems. The "baby Bells" are spending billions to team up with cable companies. Even the cable companies are now adding telephone to their video services. Just the other day, the media and Wall Street went crazy over one regional Bell's new announcement that it was getting into the movie production business. Yeah, the FCC is really pushing their "dial-up video" promotion to have competition in the cable business. They want competition in all of the local telecommunications services. So, open it up so anyone can be a local telephone company! In the old days, New York City had 29 different telephone companies. Yes, they did have problems where people using different systems couldn't talk to each other, but we've solved all those kinds of problems today. Telecommunications is big business now. It is wide open to mergers, acquisitions, and takeovers.
>
> So it took a long time to become what it is today. This is the late 1990s. The regional Bell operating companies (RBOCs) have been dragging their heels on opening up the local loop—the last mile. They are a real bottleneck! What America needs is more competition. So open it up. Now! What do you mean the rural, too? Who cares about the rural areas today? Only 3% of the people live in the very rural communities. The real opportunity is in the big cities. What we need are these new superhighways so everyone, anyone can drive their own type of vehicle on it at any speed. It can grow and grow just like Internet did. It really doesn't matter what it is, or how it is initially constructed. It can change and become anything we want it to be. So, let's begin with entertainment. There is big money in entertainment. There are billions in dial-up movies, shopping at home, and games applications. Just put a box on top of your television, and you are on the information highway, and away you go!
>
> What? What do you mean that everything I just said is all hype and hoopla? What did you say? That Internet is not a robust, secure, stable network? You heard that it is just a collection of service nodes connected

by leased trunks and accessed by remote users from dial-up low-speed voice-grade modems through the voice network? Schools obtain access to it via their local area networks? Yes, I believe that's basically true. Yes, I agree that we do need universal data transportation, and that there are all forms of transportation. For example, there is a big difference between a bicycle built for two and a Concord jet . . . Of course, but at least they are both transportation. You have to begin someplace. Okay, okay, I didn't really mean to say that Internet could do anything or that we can evolve from any starting place to anything we might want. Yes, if we really want all those robust, secure, private data-handling features, if it really will be moving all forms of data to all those different types of databases, we really need to start over with something different. I now get up at 3:00 in the morning to have better access to Internet transport, and if the whole society really does begin to use it, it probably will just break down; but Internet did prove the case that the world wants a data network that enables interconnection of addressable data terminals. How well it interconnects them and how good the connection is, especially as more and more users want to be connected, is another story. Sort of like the problems in the dial-up fax network.

Well, what about the cable company's coax tree-structured network versus fiber? Yes, the fiber does have tremendous capabilities. Yes, a fully switched fiber network enabling fully interactive videophone is quite different from broadcast single-direction entertainment over a coax cable system. So, maybe I did exaggerate when I said the cable firm's coax network was fully interactive, but scientists and engineers are working on a limited upstream path to enable some forms of data, voice, and video interexchange. I do hear that it is quite limited and complex, especially as they try to insert the appropriate switching nodes in hybrid coax-fiber facilities. I do agree that it does seem a little foolish to plow up the streets and put in coax when the fiber today is capable of moving billions and billions of bits, and coax only millions. I have heard that the real cost is in digging up the streets. Yes, I admit that most businesses don't have television sets in their offices. They do need business data networks for their computers, much more than Internet can provide. You realize Internet really just proved the need for having the capability to interexchange data messages. It sort of grew and grew until gruesome. Similarly, LANs simply provided an internal connection bus for extending the local mainframe computer throughout the building, since the initial PBXs ignored the data users' data transport needs. Later they were extended to interconnect more dispersed processors and terminals. But nothing else was offered to us from the RBOCs. What else could we do?

What do you mean that connecting LANs by point-to-point high-speed pipes is not the final answer for business? Oh, I see. The next step is a fully switched, addressable, variable-bandwidth network, having alternate paths and routing capabilities to ensure survivability, and special handling to ensure error-free, secure movement of information with tightly monitored network traffic handling and congestion-control management systems. Yes, this is quite different from Internet or government-sponsored information network pipes, but do we really need it? Why not just give everyone a modem and let them use the dial-up network, and use these leased facilities or government-supported facilities when necessary? Yes, I realize the voice network was designed for voice traffic, and as we add more and more data traffic, we are abusing it and limiting its voice-handling capabilities. I have heard where firms on Monday morning dial across town and keep dialing until they get a good line capable of moving the data and then leave it up for a week. Of course, the networks weren't designed for this type of usage. Yes, I realize there are certain circumstances where it is important to ensure the right type of data arrives properly, especially when the applications are quite personal. Oh, when we look at it that way, yes, if during my operation the hospital is routing my X-rays to a specialist in the Mayo clinic for consultation, I do want them to get them quickly and accurately.

Well, now that I think about it, I guess we do need to look again at all this hype by Wall Street. There is more at stake than a company's stock going up on a media announcement or government open-arena recommendation. I guess we really need to step back and ask how we are really going to achieve a fully interactive, variable-bandwidth information network.

So, what is the missing infrastructure? We need to determine what type of infrastructure is required for what type of information movement to sustain and encourage growth and usage. We need to know how we can best handle data so that these parallel data and video networks work as well as our current voice network and still enable local-area competition so we do achieve an open arena, but one that can sustain its openness and won't go down, one that does enable appropriate private-to-public internetworking interconnections. Yes, we do need to determine how wireless and wireline services work together.

Yes, I agree we should remember back a few years, nearly 100 years or so (1889), when "Father A. B. Strowger," the friendly undertaker in LePorte, Indiana decided that there were too many people in town waiting to be connected by the local operator's manual connection panel. So he invented the step-by-step electromechanical switch so "Ma Bell" could automate the connectivity process. Thus the world of voice switching systems was born. So, I do realize that as speed, universal addressing, performance, and throughput become more and more an issue with the Internet and LANs, the search will continue for finding a better way, a more versatile way, by constructing special data-handling public networks to meet expanding data user needs. Yes, it is time to stop and rethink the basic questions: What's happening? Where are we going? Where do we want to go? How can we get there? I guess we shouldn't dump billions and then find that we built the wrong thing or that it falls apart attempting to do a more massive job than we originally intended. Let's stop right now and take time to think this through.

As we look at the full spectrum of exciting future information-services opportunities that will be provided by the unrelenting advancement of new technical possibilities, we need to pause and ask ourselves: What is our business? Why are we in business? What is it that we wish to achieve? In descending order of importance, is it not the following?

- Create value for society, i.e., make society "better."
- Create value so customers obtain quality products and services.
- Create employment for people who share this common vision and these goals.
- Create long-term stockholder value.
- Create a viable business to pass on to the next generation.

There is a need for a clear, crisp, detailed vision that provides the complete picture of not only where we are going short-term, but also where we need to end up long-term. As we move forward into the information era, this analysis attempts to properly bring this needed vision of the future information marketplace into proper focus in order to establish a logical direction and help eliminate current confusion and conflicts.

As we consider the new information road to the new information marketplace, before we attempt to build it, we need to ask ourselves: What exactly is an information highway? Why should we consider it? How much will it cost? Who will pay for it? When? Where? In what time period? Where does it go? To what type of new marketplace? For whom? For what type of service vehicles?

A market view—an application view

First and foremost, we need to ask ourselves: How does an information highway make life a little easier, a little better? Will its impact on society improve our quality of life, or will we be worse off than we are today? What are the applications? It is extremely important to examine not only our current mode of operation, but also potential future modes in different market sectors to see how the information highway can enable us to be most effective.

For example, the express delivery services that blossomed during the 1980s have physically transferred large volumes of information in the 1990s. Local, national, and international mail carriers emulate the Federal Express "next-day" delivery system, and we see tremendous resources, both human and mechanical, put into motion to meet these highly challenging, fast delivery goals. Pickup/delivery vehicles scurry here and there throughout the day while jets travel throughout the night to sorting areas. Can society continue to consume large amounts of fuel and congest the highways and byways?

As firms become more physically distributed locally, nationally, and globally, and businessmen and women hasten to attend half-day meetings in distant conferences, businesses are searching for a more economical and less time-consuming overall solution. As this problem becomes more challenging, complicated, and universal, no single company will have all the communication resources available for providing an adequate solution. There is a need for an economical, publicly shared communications solution that reduces cost and ensures expanded availability.

As general practitioners have required more and more assistance from specialized associates who might not be locally available, health care has become more complicated. One specialist at the Mayo Clinic noted that his time effectiveness and scope of involvement could be drastically improved if sufficient telecommunications facilities were readily and economically available to enable him to participate effectively in remote surgery, patient diagnosis, test analysis, and conference discussions.

Similarly, communications can greatly facilitate the "just-in-time" arrival of needed assembly-line parts from both local and remote distributors. This greatly reduces the final cost of the manufactured product, ensures more global competitiveness, and enhances the products' chances for greater penetration into mass markets.

Education is another key need that remote communication facilities can successfully address. Students can not only attend distant classrooms, but also obtain specialized briefings provided from video storage devices containing the views of leaders in their particular field of endeavor. If accessible from the home as well as the office and university, not only will doctors, engineers, and others obtain selected continuing education, but also individual students from prekindergarten through college can have better access to better education.

With increased "work-at-home" capabilities, more people will be able to be closer to their children and still be able to actively participate to varying desired degrees in the business community. Similarly, small businesses can successfully operate from the home more effectively and efficiently.

With the technological advances that enable instantaneous access to remote databases, small businesses can be global participants as they become both international suppliers and users of specialized client-server services.

This list of applications goes on and on. Real-estate listings remotely show home videos; doctors at home or at weekend retreats review X-rays; remote doctors in small villages obtain consultation from big-city research hospitals' specialists and access to federal medical data banks; auto part stores access central warehouses for inventory; supermarket cashiers perform point-of-sale transactions (not only logging outgoing inventory but also performing instantaneous debit-card money transfers); local voters express their desires on particular bills to their congressional representatives in Washington; and the democratic process opens up new avenues for the sharing of information to ensure that the common good is preserved and maintained.

A technical view

With the continuing advancement of technology comes new challenges and opportunities, as well as new competition. These differing aspects must be fully considered as new products and services are addressed. In this regard, the current and future technological advancements can be separated into an exciting array of narrowband-wideband-broadband offerings in which broadcast television is simply a subset of both wireless and wireline broadband services. As seen in applications such as entertainment, the communications and computer industries have become more and more integrated. Their boundaries are becoming more and more blurred, such that we can no longer tolerate expenditures here and there in an independent, haphazard manner. Programs and projects should provide products and services that become building blocks that establish a strong infrastructure to provide a supporting bridge to the future. Hence, there is a need to describe the narrowband, wideband, and broadband (N-W-B) technical possibilities and market opportunities for both wireline and wireless voice, data, and video services.

It is important to show that the information highway is much more than just a broadcast entertainment highway or a broadband transport (nonswitched) superhighway for connecting research-and-development (R&D) institutions and universities. The information highway must become a fully interactive, switched broadband transport that also consists of a narrowband ISDN copper-based highway for small businesses, as well as a fully switched wideband road for larger businesses. It is also essential to understand that glass fiber has infinite possibilities for interactive terabit (thousand-billion bit) traffic, while coax is only an interim megabit (million bit) broadcast technology. Hence, the need for a full, panoramic view of the future and a fully detailed assessment of the particular programs and projects that address the challenge of achieving success in the forthcoming information marketplace.

As we enter this final phase of this millennium, there has never been a greater opportunity to achieve an exciting new information age. Now we can readily see that technology has reached the "deployment" state; no public, medium-speed narrowband data network exists today at 64,000/128,000-bits-per-second (bps) rates. [It takes 8 bits (called a byte) to represent an alphanumeric character such as the letter "A" or the number "8;" so 64,000 or 64K (K = 1,000) bits are effectively 8000 characters, which at 4/5 characters per word is approximately 2000 words per second. Compare this to current voice-grade rates of 4800 bits per second or approximately 100 words per second. This achieves transport capabilities that are 20 times faster than voice-grade dial-up transports in use today]. Similarly, no fully switched public wideband data network exists today transporting multiples of 64,000 bits per second ($n \times$ 64-Kbps), 1,540,000-bps (1.54-Mbps), 6.3-Mbps, or 45-Mbps rates; no public, switched broadband data network exists today at 51-

Mbps, 155-Mbps, or 600-Mbps rates; large businesses have only seen private networking as their solution for multiunit interfirm communication, but they are now asking for public shared data transport solutions; small businesses are just awakening to their expanding business needs for internetworking personal computers (PCs) and obtaining access to remote databases; current copper plant can easily support narrowband ISDN, both basic 64/128-Kbps and primary 1.5-Mbps rates; existing plant can be selectively upgraded to provide switched wideband and first-stage broadband offerings, up to 45 Mbps; switched broadband can be initially established on an overlay basis for selected offerings as standards and next-generation switches become available; transport data is in an exponential growth phase during the 1990s; data occurs in both the bursty and the continuous-bit-rate mode of operation, requiring different types of data transport; internetworking private networks via the public network is a key opportunity; applications exist today for automating information exchange for each of the major 18 industry groups, both within and across their communities of interest.

Hence, as we look at the information road, we need to differentiate it in terms of these N-W-B considerations, as well as its various types of media that facilitate its operation. For example: What about the current multiple-frequency analog transport-based copper plant? What new capabilities are offered when these signals are digitized? What value is added as information is quantized and digitized into on/off pulse streams (bits) that can then be packaged into various streams of data channels that are transported from point to point or switched from many points to many points? What is the advantage of using even more sophisticated mechanisms where information is packetized and interlaced with other messages, similar to integrating various chains of boxcar-type vehicles using store-and-forward types of systems. Hence, the type of delivery medium greatly enhances and changes the final service, similar to the differing effects caused by the various types of transportation delivery vehicles: a bicycle, a car, a train, a ship, or an airplane. Even these separate entities are further differentiated by technology. For example: Is it a prop airplane or a jet airplane? Airplanes can again be broken down further by technological advances such as the differences between a DC3, a Boeing 747, or a Concord.

A broader perspective: MARK-TEC

The overall service changes substantially, depending on the technology used in the delivery vehicles traveling over the various types of information highways. We need to define both the type of roads and the type of vehicles in terms of their capacity, transport rates, error rates (degree of bumpiness), entrance ramps, lane control mechanisms, checkpoints, etc. Similarly, we need to consider the drivers' desires and capabilities. Putting the senior citizen in a high-powered race car is usually as undesirable as putting a baby's tricycle on a four-lane speedway. We need to match the driver with the vehicle and the roadway. For example, we need to understand the differences among the overlay/bypass networks that enable direct access to interexchange carriers (IXCs). We need to know what they do, how they work, and where their access should be located. These alternatives need to be understood in advance and become part of the overall private-to-public networking aspects of the provision of layered networks' layered services from multiple providers.

As more and more information players awaken to the new opportunities and rise to the occasion, we'll see more and more competition via private fiber overlay networks; shared switched broadband service networks; local-global value added networks (VANs); private specialized service consortiums; community-of-interest market servers, such as medical; competitive access providers' (CAPs') wideband switching hubs for networking private networks; interexchange carriers (IXCs)—merging with computer firms that also offer local overlay networks to specialized database services.

Regarding regulatory issues, technological advances will enable transport bandwidth to become a commodity. We will see concerns for sharing a competitive marketplace shift focus from transport networks to application services, and the Federal Communications Commission (FCC) will continue its drive to ensure a multiplayer arena. Each state will hasten to ensure that it is considered a global habitant in the global information village.

With these new opportunities, new competition, and new regulatory/nonregulatory considerations and restrictions come new challenges. Now is the time to address the following changes and needs: major changes in the local exchange carriers' (LECs') mode of operation (traditionally voice-network oriented) in order to build a ubiquitous, full-featured public data network; a major bandwidth repricing policy in order to price for growth and price on a service basis; a last-mile fiber deployment policy; unregulated/regulated separation of future data and video services; data technology opportunity training for executives, management, and work force; new competitive strategies to address new competition in the local arena; long-term versus short-term revenue plans; long-term financing strategies; new emphasis on repositioning regional Bell operating companies (RBOCs) and other local exchange carriers (independents) in the public switched-data transport arena.

Finally, let's look at the toll ways, toll bridges, and toll gates. Some have their purpose, but many unreasonably limit the progress and growth of traffic, causing huge bottlenecks and blockages as traffic movement is limited, thereby encouraging economic bypass. Pricing algorithms that foster growth are key in encouraging and controlling the continuous, orderly flow of traffic. We can no longer price services artificially high and then wonder why customer acceptance is minimal. If indeed we are in a public, mass-market information business, we need to design all aspects of the offering accordingly. Hence, the "information road" leading to a new age, a new era, the information era, was conceived and defined in 1984 at the time of divestiture by my earlier works (*Telecommunications Management Planning: Networks, Products, & Services*, McGraw-Hill, 1987; *ISDN in the Information Marketplace*, McGraw-Hill, 1989). Those books denoted a much greater opportunity than that indicated by the newly coined 1994 Washington term, "superhighway." In reality, it is an N-W-B information highway leading to an exciting, far-reaching global information marketplace.

As we strive to achieve an exciting information-based society in which our children, and their children, can flourish and grow to obtain new heights of civilization, each piece of the puzzle must fit into the overall picture or be discarded. We need to search for just the right piece to successfully achieve the new "information-based" society, the type of society that we wish to leave behind as a stepping stone so that the next millennium is properly launched with its endless possibilities for generations yet to come. To accomplish these objectives, advance, front-end planning is essential. We need not only a clear vision for the future, but also the functional plans necessary to achieve it—both a plan for the future and a plan of action required to achieve these endeavors. This results in several strategies for achieving an exciting array of future networks, products, and services. These opportunities are identified herein at the conceptual what, why, how, and for-whom level. From there, appropriate marketing-technical (MARK-TEC) program and project management teams can perform the necessary product-definition planning to ensure that all proper and essential aspects of these offerings have been carefully considered in order to obtain executive understanding, buy-in, commitment, and support.

One leading information network provider's formally stated vision is that, "In the year 2000, we will be the finest company in the world in connecting people with their world." We need to establish a plan of action for achieving this vision by focusing on a new, far-reaching, challenging goal: "To establish a new revenue-producing public data information network, providing narrowband, wideband, and broadband services in timely, phased offerings." To accomplish this, we need to translate this vision into more detailed strategies, objectives, plans, programs, projects, and organizations that deliver the necessary networks, products, and services required to address diverse customer information applications. So let's take a more expanded view and consider all aspects, both marketing and technical, to formulate a better understanding. We need to attempt to answer the provocative and thought-provoking questions: Where do we want to go? How can we get there? Will we like it once we arrive?

Billions will be spent during the upcoming decades as we deploy various new networks. How this money will be spent and how each building block comes together must be determined by a clear vision of the future.

Thus, now is the time to step back and reflect on how to successfully achieve the necessary information infrastructure to deploy fully interactive, N-W-B services ubiquitously from coast to coast, ocean to ocean, and nation to nation in the forthcoming global information marketplace.

The bottom line

As we look forward to the technical possibilities and market opportunities of the future global information millennium, let's consider: What does all this mean? What impact will future telecommunications have on our lifestyles, our daily lives, our mode of operation as we perform our tasks, our businesses, our health, our homes, our children, their children? What could be, should be, would be, might be, will be?

During the past decades, we have seen tremendous economic shifts that have greatly affected ourselves and our families. Incomes for middle-class males have dropped 23%. Wives no longer work to obtain the luxuries; they work to help pay the basics for survival. A $30,000 home built in the Midwest in 1964 costs $120,000 in 1994, while an equivalent new home costs $200,000. This same home on the East or West Coast of the United states could cost from $250,000 to $300,000, up a factor of 10. Most other costs of various items are up a factor of 6–7 over this same time period. Unfortunately, long-term interest rates of 5½% in the 1960s skyrocketed to 13–15% in the late 1970s and leveled off after a brief dip to 9½% to 10½% in the 1990s. Similarly, a $9–10,000 income in the early 1960s is equivalent to a $50,000 income in the 1990s in terms of purchasing power. The gross national product of the United States in the late 1960s was 75% of the world GNP, while in the mid 1990s it reached only 23%. During the 1985–1995 timeframe, not one of the Fortune 500 firms added a new job. In contrast, there has been a painful trail of forced early retirements, layoffs, downsizing, rightsizing, and reengineering to reduce costs and make the firms more competitive.

On the personal side, day-care costs in the mid 1990s for two children are around $800 per month, or $12,000 per year pretax, while latchkey kids, gangs, high divorce rates, struggling single parents, etc. have created tremendous pressure in the home and on the family. Even the busy super-moms are suffering tremendous pain as they attempt to cope with both business and home issues. As they become more and more successful in the business community and attempt to hold the family together, their positions require extensive travel as society becomes more global.

In turn, the computer revolution has indeed begun to take place in earnest in the 1990s, as global competition drives firms to new processes that speed up product delivery, just-in-time product inventory, better-quality management, and employee participation in the process of obtaining workplace improvements. In the industrial countries, populations have become older and more affluent, with 60% of government expenditures for the elderly; at the same time, the third world has become younger, with 50% of the populations of developing nations under 18. There are many new needs in the areas of health care and new job creation, especially as computer automation and high-technology solutions require a higher skill level for all workers, from management to assembly line.

Similarly, the high concentration of population in the urban environments takes its toll as cities grow from metropolises to megalopolises to mega-megalopolises, where parkways, freeways, tollways, overpasses, and bypasses become clogged with congestion during rush hours and extend congestion periods later and later, earlier and earlier. For example, in Washington, D.C. rush hour begins at 3:30 p.m. and continues to 7:30–8:00 p.m., while early-morning congestion begins at 6:00 a.m. and lasts until 10:00 a.m. Similarly, flex hours in corporate firms in Stamford, Connecticut encourage differing starting hours to enable workers to more easily travel over the Merrit Parkway; the New York commute trip from Fairfield County is now 2½ hours each way. Even Seattle's coastal highways are now jammed due to heavy influxes in the early 1980s. In the 1970s a driver could actually stop an automobile at certain times of the day on the (then) more remote freeways without fear of damage. Even Denver's congestion has exploded due to growth from Californians in the mid-1990s. Chicago's western suburbs, and, of course, L.A.'s freeway parking lots reached the crisis state many years ago. They now have total rush-hour grid lock.

As populations have moved to the urban communities and the rural villages have disappeared, world populations now live on only 1% of the land, with only 3% in rural communities. This causes urban housing to become more and more scarce and expensive. There is a crucial need for social change to reduce these economic and personal pressures, help restore a lost quality of life, and then hopefully to improve upon it.

So why build a new information infrastructure? Why pursue a long-term program when the financial community continues to resist long-term investments over 6–9 years without seeing a reasonable early return on their investment? In the past, only the Department of Defense took on long-term programs. For example, the interstate highway across America was originally begun under a defense highway bill. Today, Wall Street would never have supported the original cost for constructing the current universal voice telephone network. It is more simple to build overlay networks serving private communities where customers are already lined up, where costs and revenue streams can be more easily identified and obtained. Building robust, secure, survivable networks is truly a challenge, much more than connecting a few tin cans together with a string. This might have been okay for two neighborhood children, but when the communications network encompasses not a couple, not a hundred, not a thousand, not a hundred thousand, not a million, but hundreds of millions of users requiring interconnections for billions of calls, this is indeed a complex and challenging program requiring long-term commitment.

So as we look to the future, why not simply deploy a network for more television channels or let data calls try to get through on the dial-up voice network? Or again, simply construct a simple overlay network with all its lack of security and congestion problems, such as Internet, which will in time simply collapse upon its own success. So what is the incentive? What are the applications? What are the needs that have to be fulfilled, and are we willing to do what's necessary to fulfill them? What will life be like in the new millennium? How can we address and resolve our personal economic problems, reduce our stress, become more productive, and provide competitive products and services in the forthcoming global information society?

Observations and conclusions

For one thing, many people will be able to work at home and live in a more rural location. This will limit the need for day care, reduce economic pressures on the family, and diminish the influence of after-school gangs.

Using the electronic broadband network, the heart specialist at the Mayo Clinic will be able to remotely participate in several complex operations in the same day, thousands of miles apart in opposite directions from the clinic. Radiologists will be able to review X-rays at 2:00 a.m. at home in their studies, without having to rush to their clinics or hospitals, where delay could mean life or death.

Background checks will be quickly and accurately obtained for an applicant desiring a responsible teaching position in a local school. Investors will be able to instantaneously electronically access their financial investments to determine their actual worth.

Education will not only be broadcast to remote areas so viewers can see the professor and the blackboard (or electronic board), but the professor will also have the ability to selectively see the students to assess comprehension and field questions.

Not only will gasoline stations perform credit checks and debit transactions directly from bank accounts, but also the amount of gasoline sold will determine deployment schedules for delivery trucks to refill the station's holding tanks and determine local depot needs for the entire service area. This applies to all fast-food chains as well, where 1:00 a.m. truck deliveries would be determined by the amount and type of daily sales.

Home shopping will never replace going to stores for touch and feel, but it can certainly aid the prospective buyers in their search for the right product at the right sale price. Shoppers always hate to find out the next day that an item was on sale at half the price down the street.

Remote work places, electronically tied to central offices, will enable shorter work distances, less-tired commutes, and free parking. This will help to eliminate or reduce one of our two major daily jobs: getting to work and actually doing the work.

Electronic library search mechanisms will offer instantaneous access to information, independent of current location or distance; this will provide information from anywhere at anytime at our fingertips. We'll "let our fingers do the walking."

Inventory control mechanisms will enable chain stores to access central warehouses to obtain specific items. Doctors and responsible officials will have the ability to search federal data banks such as poison or disease control centers. Anyone having video conversations on different networks with different capabilities, perhaps in different countries, will be able to intercommunicate together. Remote villagers will obtain access to urban culture in distant lands.

As we pursue the future information marketplace, it is essential to add reality to these endeavors by providing perspective users with answers to the following questions: When will a public data network be universally available and specifically structured to transport volumes and volumes of data in an error-free mode at least 20 to 40 times faster than the existing voice-grade network, such as at narrowband ISDN rates of 128,000 bits per second?

Where will this type of data network be available? Only in the major cities? In the business districts? In the suburban markets? In the rural communities? In the homes? On the farms? From ocean to ocean?

Will the network have a data user's directory, with an address for every terminal device, just like the telephone directory? Will it have all the features of a robust data-handling network (such as those defined for the world communication networks by the International Telecommunication Union for both circuit switching (individual paths) and packet switching (shared paths) at narrowband ISDN rates of 128,000 bits per second)?

Will the network be priced at reasonable rates to encourage universal usage? Will the network have full-feature functionality within the local calling area, as well as across the state, region, nation, and globe? When will the fully switched, addressable high-speed networks become available (first for megabit rates of 1.5 million to 45 million bits per second, and then for broadband ISDN multimegabit rates of 155 million and 600 million bits per second, requiring fiber to the home and office)?

When will users be able to have fully switched video conversations covering the full range of speeds from narrowband to broadband so that anyone can see anyone anytime over addressable, fully switched, fully interactive facilities? Where will these high-speed services be available? Will they support fully switched integrated multimedia voice, data, text, image, graphics, and video offerings without congestion, blockage, contention, or outage for millions of users, and not just for a few limited point-to-point users as private overlays?

Will the network be ubiquitously available and supported with appropriate network management, survivability, security, and privacy mechanisms? Will the network have the availability of the traditional telephone network? Will it be based on sophisticated switching, addressing, and transport hierarchical structures, using robust technologies with supporting operational capabilities that handle extensive universal growth and usage, or will offerings be implemented on an overlay ad hoc basis using relatively limited technologies? For example, will offerings use heavy-load, public, industrial-grade system structures or those tailored to private customer-premise types of systems? Will the products be developed for highly cost-competitive markets by firms using the skills of only a few designers, or will they be developed by more sophisticated laboratories employing numerous skills to obtain a highly robust, quality product? When will fiber be deployed in the urban, suburban, and rural communities, not on a shared, contention, constrained basis, but with the full range of transport services instantaneously available for each user in the home or office? What can we order today, tomorrow, and in the not so distant future? When and where? To what extent? To what degree?

There is a substantial difference between a broadcast medium with an asymmetrical, large, one-directional path and a small return path versus a symmetrical, equal, fully switched, two-way offering. There are substantial differences among bits, kilobits, megabits, gigabits, and terabits technologies. Each has its own capabilities and limitations. Long-term, robust offerings that provide an exciting array of features and services compete at the surface level with short-term technologies that offer limited, quick, cheap solutions. So no matter where and when these offerings are available, it is essential to understand both the what and the how of their capabilities. These are indeed convoluted issues as we attempt to obtain the right network with the right services at the right location at the right time.

The bottom line is that life can become substantially different, substantially improved, and hopefully substantially less expensive. We can reduce economic pressures and stress levels if indeed the correct, needed infrastructure is universally available. So what is the vision? What are the plans? What is the plan of action for achieving this missing, supporting infrastructure locally, nationally, and internationally? Now is the time to construct this structure—not some time in the distant future.

Although this analysis and assessment are personal views and do not necessarily reflect those of any affiliated organizations, they are offered to provide a timely, appropriate platform to stimulate and support further thinking and dialogue among those participating in this exciting information arena.

Hopefully, when our children's children look back over the twenty-first century, they will see that we indeed effectively placed the appropriate

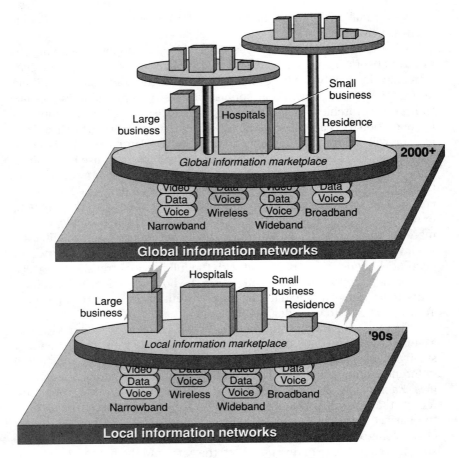

Fig. I-1. The information society.

building blocks to properly support the global information society in the third millennium, the information millennium. (See Fig. I-1.)

Now is the time.
The time is now.
Now never waits.

Part 1

Outlook for the future

"There is a future in this great world
for those with the will to work
and the imagination to see it."

Louis L'Amour

1
Vision of the future

"We are such stuff
as dreams are made of . . .
and our little life
is rounded with a sleep."
William Shakespeare

As the telecommunications industry looks forward to its one-hundred-twenty-fifth-year anniversary some time in the early phase of the information millennium, it is essential that we pause and assess not only where we are going, but also where we have been and how we achieved our present position. For only in reviewing our somewhat evolutionary path, noting its twists and turns, can we assess our strengths and weaknesses and establish realistic goals for our new endeavors.

Ten years after divestiture, the telecommunications marketplace awakened to a full range of "information-handling" service opportunities: origination, transfer, storage, access, processing, manipulation, and presentation. Traditionally, the common carriers, both local exchange (LEC) and interexchange (IXC), had been preoccupied with establishing a switched network that was extensively automated to handle voice calls. With this mission, the core business was made universally and ubiquitously available and affordable via a highly efficient and effective operation. Over the years, the "black" phone was replaced by numerous varieties of different shapes and colors. Former manual "operator-please" services are now established directly by customer dialing. Direct dial access has become global. Services are customized, and data transport bandwidth has become greater, shared, and less expensive.

By the mid-1990s, competition had entered every segment of the marketplace, vying for every form of customer communication: voice, data, image, text, graphic, or video over both the wireline and wireless media. Unfortunately, or fortunately, depending on one's perspective, data user needs and expectations were mainly ignored in the 1960s, 1970s, and 1980s, and these customers turned to private networks to address their data transfer requirements.

Local area networks (LANs) were established on customer premises to extend the internal back-plane bus of computer mainframes throughout the external local user environment. This enabled users to send and receive information at somewhat respectable speeds over existing copper and coaxial facilities. Then the new "computer-on-a-chip" microelectronics technologies enabled smaller, highly versatile computers to participate in every facet of the operation of every major industry, from health care to education, thereby helping to automate all internal tasks. These local internal networks were deployed to enable numerous new families of peripheral systems to exchange information with sophisticated, large mainframes and achieve access to huge central databases. (See Fig. 1-1.)

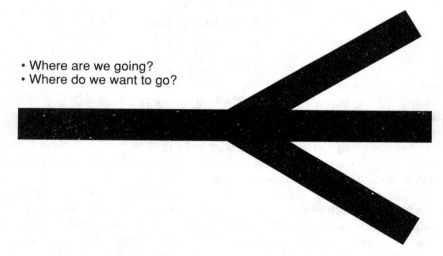

• Where are we going?
• Where do we want to go?

Fig. 1-1. Vision for the future.

In this, the "many-to-one," "one-to-many" data networks were established, allowing remote stations to exchange information with more and more powerful supercomputers. This then established the need and paved the way for wide area networks (WANs) and metropolitan area networks (MANs) to extend the reach of these supersystems to more and more disparate terminals located in any industry throughout the community.

As evolution would show, the next step in "computer networking" was to enable different terminals to talk to different mainframes. Protocol converters enabled disparate systems of any size, shape, or form to interrelate. Some believed this would be the destiny of computers, as processing power and versatility doubled and doubled again approximately every two years or so throughout the 1980s.

By the early 1990s, it was quite apparent that other forces were at work, as integrated-circuit manufacturers developed workstations having very high density chips, achieving the power and capabilities of earlier mainframes. Data creation, manipulation, and processing could now take place at these remote satellite workstations, which required less and less communication with single mainframes but more and more communication with numerous distributed data storage and processing systems that were located locally, nationally, and internationally. These personal computers (PCs) and knowledge workstations (WSs) have advanced during the 1990s to become quite sophisticated. They enable multimedia communication, but no longer among "many-to-one" or "one-to-many" applications, but for "many-to-many" or "any-to-any!"

Hence, the need is readily apparent for establishing public data networks that traverse the spectrum of narrowband, wideband, and broadband offerings. "Data handling" in the narrowband world becomes more and more "information handling," as its breadth and scope is advanced in the wideband and broadband domain. There, customers are provided with switched bandwidth facilities that enable them to dynamically, as needed, transfer "n" number of channels of information to any location.

As initial bursty variable-bit-rate (VBR) dataphone and imagephone traffic gives way to more and more continuous-bit-rate (CBR) videophone and viewphone usage, it is essential that packet-switching facilities are augmented with circuit-switching capabilities. ("Phone" is used here in the generic sense in that the telephone is on an audio information-handling terminal and the dataphone is on a data information-handling terminal.) This will enable the availability of expanding bandwidth, provided by photonic switching systems, to meet increasing market demands. Hence, not only must fiber transport capabilities expand with expanding video usage, but switching capabilities must also expand and adapt with changing information traffic attributes as users shift from voice to data, to image, to video, to view. To deploy a few fibers to handle all voice traffic needs will not address the facility loads generated by high-definition videophone calls. Similarly, to deploy a broadcast facility to offer a few conventional television channels does not address the possibilities of providing hundreds of full, high-definition, high-resolution offerings in the twenty-first century. Switched fiber facilities will rise to the challenge, especially as radio spectrum becomes scarce.

However, physical balance is needed. High expectations require high expenditures to achieve ubiquitous fiber-based offerings. To be achievable and realistic, fiber deployment plans must be phased in over a 20-year pro-

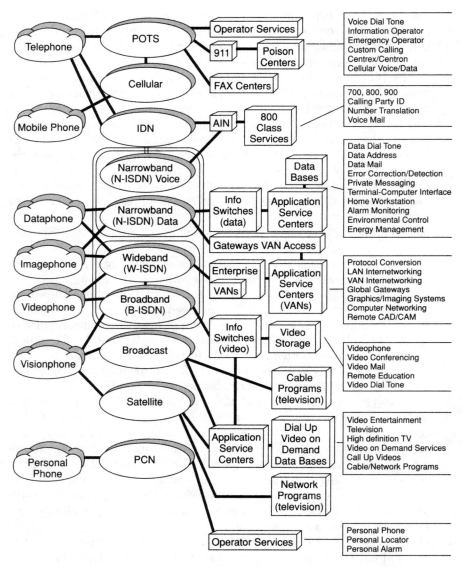

Fig. 1-2. Information telecommunications vision of the future.

gram. We can maximize the use of existing copper plant during the transition phases by offering narrowband switched data services. As we view the future, there is a shift from the traditional POTS (plain old telephone services) network to integrated digital networks (IDN). (See Fig. 1-2.) These will form the basis for shared voice and data movement in the digital mode over copper facilities using the ISDN (Integrated Services Digital Network) format. This provides new public "data-handling" highways in both circuit- and packet-switching form at speeds of 64,000 and 128,000 bits per second. These dataphone, fully addressable offerings can be then expanded to provide "image-handling" capabilities, which are then augmented by more bandwidth capabilities, as switched wideband systems become available at speeds up to 45 million bits per second in the early stages of fiber deployment. (These wideband systems will use backward-compatible virtual tributory SONET transport signaling STDs.) Then, first selectively, and finally ubiquitously, full videophone/viewphone switched SONET-based broadband facilities can be overlaid with broadcast capabilities, providing not only eye-to-eye, person-to-person, high-quality communication, but also television and selective high-definition, high-resolution educational and entertainment programs. Supplementing this array of wireline offerings are the wireless cellular and the personal phone capabilities that enable anyone to be reached anywhere, anyplace, anytime.

This completes the "vision of the future," showing the "network-of-networks" picture of potential telecommunication technical possibilities. Once these information highways are established into the information marketplace, their information services will flourish, as new singular and shared, integrated and autonomous service centers are equipped to offer the full spectrum of information services. Indeed, by the turn of the century, 125 years after the invention of the telephone and having seen the formation and subsequent breakup of the Bell System, we will have come a long way; but we are only at the beginning of a new path, a new highway, an information highway. Hopefully, it will lead us to more and more exciting possibilities and opportunities that will help us address our changing, complex societal needs in a new century, the information century, in a new millennium, the information millennium.

2

Narrowband information

*"One of the pleasantest things in the world
is going on a journey . . .
Give me the clear blue sky above,
the green turf beneath my feet,
a narrow winding road before me,
—on then to thinking . . .
It is hard if I cannot start some game
on these lone heaths."*

William Hazlitt

The existing telecommunications network is established over a copper twisted-pair wireline facility. It was originally designed to transport voice conversations in the 4000-cycle analog frequency range. With the advent of new technology, the telephone company's multihundred-billion-dollar plant is being transformed into a voice, data, and video information-handling facility. As analog frequencies are changed into digital on/off pulses, service providers are using these new technologies to leverage their initial asset from being simply a voice network to obtaining a parallel public digital data network.

A narrowband public digital data network

As we address these new data-handling opportunities, we see many questions that must be addressed: Exactly what is a public data network and who is it for? What is N-ISDN and what new revenue streams does it offer? What are the building-block products? What features do they provide? What data and video services can be offered? Where? When? How are they provisioned and supported? What customer needs are met? For what tasks? In what applications? Exactly how will dataphone and videophone be delivered, advertised, sold, and supported? With what changes to the network, market distribution channels, regulatory tariffs, and public policy positions?

What are the network's chief benefits? Who are the benefactors? What is its impact on society? What is its impact on network providers and suppliers? How can the local exchange carriers foster enhanced data services by overlaying service nodes? What services will be provided in these service nodes? Application service centers? What open network architecture (ONA) standards will be needed for advanced intelligent networks (AIN) and information network architectures (INA)? What data-handling customer-provided equipment (CPE) is needed? What will be the different revenues from differing types of data networks—circuit (ckt) or packet (pkt)? What handshaking/signaling is needed between CPE and the network? What data transport gateway interexchange carrier (IXC) access arrangements will be required? What interfaces to global players? How can regulators ensure that data offerings are as ubiquitous as possible? What are the new data/video transport/services' programs, projects, and products? How are they funded? With what relationships with interested suppliers? How can information telecommunications firms best organize to ensure success and commitment? Who are the data users? What are their needs?

Data users

The data users have specific needs, different from voice users, that need to be appropriately addressed and met. Data users need:

- Fast connections and fast drop-offs.
- Error-free transport.
- Assurance that data reached the desired destination successfully.
- The ability to send the data to multiple places.
- Faster and faster transport.
- A transport service that encourages usage, thereby priced for growth.
- The ability to interconnect to numerous disparate systems.
- Transport of information that is secure and confidential (private).
- The ability to block and track illegal accesses to internal systems.
- The ability to call for assistance—to converse with human operators to obtain access to remote networks, databases, and service nodes.

- The ability to transfer data files and graphic images, and narrow-band-rate interactive videophone conversations.
- A robust, survivable, economical, fully interactive, switched public data network that is ubiquitously available to addressable customer-premise "data-handling" equipment.

There are several unique information-user opportunities: telephone, dataphone, imagephone, videophone, and high-definition television. We can easily understand and relate to the voice and video offerings, but unfortunately, dataphone services are more difficult to fully appreciate. We are beginning to understand how improved data-handling networks can drastically change our current modes of operation as we substitute information for transportation vehicles. Hence, we must try to understand what these new data-handling technologies offer as we change to a world that no longer requires intolerable waiting periods to achieve accurate, timely information delivery.

A technical preview

First and foremost, it is essential to understand what exactly a public digital data network is. It is a new network, parallel to the voice network, that is designed to specifically handle data. A digital data network has substantially different technical types, attributes, and characteristics than voice. For example, a voice call is more tolerant to noise. It might be annoying, but the participants usually ignore a small background noise, or they talk louder. A voice network is designed for a certain number of voice call attempts. For the home, this is usually an average of 3 to 5 outgoing call attempts per day, whereas businesses might have 10 to 20 outgoing voice-call attempts. Small businesses such as a hardware store might have as many as 100 or so data call credit checks in a busy day. Similarly, the length of a voice call is quite different from a data call. Today, switching matrix blockages are based on mathematical voice traffic throughput (Erlang) tables, in which voice conversations have a 5- to 10-minute holding time. The present switching/transport hierarchy is constructed around a five-level model that enables efficient network management to ensure survivable and effective voice-call handling, taking advantage of analog voice-call residence-and-business traffic mixes and groupings.

With the advent of digital transport and switching, voice calls are sampled and quantized into slices, where specially assigned numbers indicate the different amplitude levels of the conversation. These numbers are represented in binary ones and zeroes (on and off pulses) called bits, where it takes 8 bits to represent each voice sample. Since the call is sampled 8000 times a second, this translates into 64,000 digital bits-per-second signals to transport each conversation. These representative bits can be combined with other digitized conversations into higher- and higher-speed digital transport streams. This enables more effective, efficient, and error-free transport as the digital signals are reconstructed and repeated as they are switched and transported about the network. What is key to this process is that once voice calls are transported in the new number-form digital mode, they can easily be combined with data (which already exists in the number form) to provide an integrated delivery system of both voice and data. This was indeed the purpose of first establishing Integrated Digital Networks (IDN): to handle digital voice calls and then add the data traffic with additional data-handling requirements to it. In this manner, the new data services were added to IDN, thereby calling it Integrated Services Digital Network (ISDN), which handles both voice and data transport. So, besides providing a digital voice-and-data-integrated access interface to the customer, it is essential to offer a switching arrangement that is designed for data calls as well as voice. That indeed was the purpose of ISDN.

Unfortunately, many did not initially understand this intent. Though it was specifically stated as a voice and data offering, it was sold as an alternative interface that enabled a second voice line service to the customer, which was used in the voice mode or to interconnect dial-up voice-grade modems operating only at low-speed rates of 4800/9600 bps. The existing voice-grade network transports data using various analog amplitude, frequency, or phase modulation/demodulation (modem) techniques over facilities having an error rate of 1 error in every 10,000 bits. This requires numerous retransmissions and acknowledgments to properly deliver the information. A typical page contains 25 lines at 10 words per line or 250 words having approximately 4 to 8 characters per word. Thus we need to transport 1000 to 1500 characters at 8 bits per character or approximately 10,000 bits per typed page.

For facsimile systems, image systems, graphic systems, and video systems, each dot or pixel is represented by 16 to 24 bits, thereby requiring 50,000 to 500,000 bits per page. Depending on quality of resolution and degree of color, these numbers might jump to a million or so bits. Of course, differing compression techniques reduce these throughput requirements; however, depending on noise conditions in the analog environment, it might take up to several minutes to satisfactorily transport images and messages, whereas in the digital environment this can be achieved in seconds. As more and more documents are transported across the varying databases, there are considerable economic savings obtained from both human and computer resources by enabling increased turnaround. One need only to watch the hours tick by as busy office staff attempt to handle the flow of low-speed faxes.

Now is the time to add the additional digital-data switching capabilities that are specifically designed to meet the new data-handling requirements and increase its effective throughput and speed capabilities to process both circuit and packet data. Packet data, where information is arranged in train-type groupings having a header (engine), message body (boxcars), and tail (caboose), can be handled in several modes of communication. In the connectionless mode, each grouping of data is transported independent of the transport path of a previous group. In the connection-oriented mode, a string of information is sent in a sequential arrangement; the first group establishes the overall path through the network, and subsequent groups of data follow over the same path to the same destination. Several users share the transport, intermixing their packetized data and thereby reducing their overall transport cost. Circuit data, where the fully switched transport connection is dedicated to the data user, available as needed, is basically handled via connection-oriented networks, similar to the process of establishing and holding a dedicated path for a voice call. Data can arrive in one of two forms: it can be bursty, containing a variable number of bits as variable-bit-rate (VBR) traffic, or it can be in the form of continuous-bit-rate (CBR) traffic as a videophone-type conversation.

Little bit, big bit

There are indeed two distinct types of data transporters. A friendly way of visualizing them might be to relate them to a caricature having somewhat human attributes. As we look at data transport systems, we see that data can occur in the bursty variable-bit-rate mode or nonbursty continuous-bit-rate mode; data can be either circuit switched or packet switched; some data calls require high transport security, others low; some data calls have a high tolerance for error, others low; in some instances, data speed is important; in other cases, it is less important; data can be sent over connection-oriented or connectionless transport mechanisms; data can be stored and forwarded, or not; extensive data internetworking might or might not be required; finally, data users can be concerned with cost, or not. Thus, there are all forms of data movement covering the entire spectrum of these choices. However, if we were to pick the two predominant types, we might describe them as type E and type I.

The type E data transporter we will call the "extrovert" type, the E bit, "the big bit." This is similar to the flashy, fast-talking type of person who is not too technical, overlooks errors, likes video services, calls direct, likes to see himself, comfortably spends lots of dollars, and eats in big bites (bytes). In actuality, this is the continuous-bit-rate, circuit-switched, high-data-speed, connection-oriented, high-error-rate-tolerance, low-transport-security, high-transport-cost data transport system.

The type I data transporter we will call the "introvert" type, the I bit, "the little bit." This is similar to the quiet person who is very technical, does not like errors, works in a cloistered, cell-type environment, is very dollar conscious, tends to save money by traveling in shared transport vehicles like buses, subways, and trains, needs to access lots of information, uses numerous networks requiring protocol conversions, and eats in small bites (bytes). In actuality, this is usually the variable-bit-rate data traffic that is bursty, packet switched, concerned with security, has a low error rate tolerance, can be stored and forwarded, requires protocol conversions for data internetworking, is less conscious with speed and more with accuracy as long as it's reasonably fast, and is thereby willing to share transport mechanisms to reduce data transport costs.

Data characteristics

So what are the characteristics of data? What network attributes are changed to specifically handle data? What do data users, the telephone company's data customers, require to assist them in the movement of their data? We have already seen many basic data network requirements in past applications—for example, in the global military data networks, business local area networks, and commercial dial-up modem-based data traffic on the public voice network. We have seen the data calls come in many shapes and sizes. However, it is essential to consider not only present, but also future applications to best appreciate the data transfer needs of the various users on the various new networks.

In so doing, we need to understand that data users generate numerous short-holding-time calls (such as credit checking), as well as long-holding-time calls for remote terminal-to-computer conversations. Data users have a low tolerance for high error rates, especially as large blocks of data are being transported and where bursts of noise cause errors that require substantial retransmissions. Users require the ability to interlace numerous data calls together between computers and terminals. Here they are insensitive to interaction delays. They need to hand off information to be transported over differing paths with differing protocols in an expedient manner, while ensuring that content integrity is maintained or at least is verifiable at distant ends. Similarly, data users arrive from different systems, each with their own internal synchronization techniques. Hence, their information arrives asynchronous to the network which, in most cases, is operating at a different transfer rate as well. This requires multiple-rate transport conversions, as well as asynchronous-to-synchronous transfer conversion mechanisms that enable private-to-public networking.

Therefore, there is a need for a new family of data-handling network parameters to ensure that the various transport access arrangements are achieved and to ensure the safe, secure, survivable movement of the data information. This is a key purpose of the public data network. It cannot go down. It must be robust. It must transport large quantities of information to every destination. It is important to understand what features enable this to be achievable and what services this form of network offers its data customers, as well as what services are provided above the network via service nodes to enhance and ensure the successful movement of data. (See Fig. 2-1.)

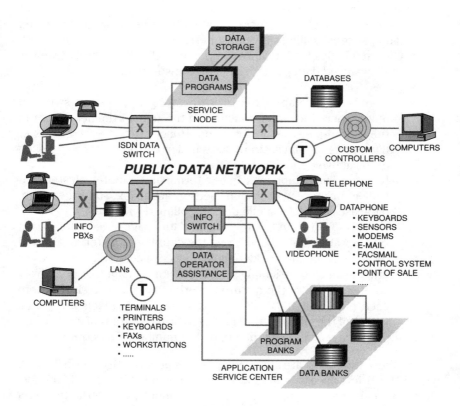

Fig. 2-1. The world of data.

Transport access arrangements

The ISDN data transfer network handles several permutations of various telephone, dataphone, imagephone, and videophone access arrangements. To do so, it has multiple transport/switching capabilities. These are identified as specific data-handling features that a public data network offers to prospective users. These capabilities are considerably different from those provided for voice-grade dial-up data modem transport. Combinations of voice and data interface arrangements are provided, thereby enabling:

- One telephone and one high-speed dataphone operating at 64,000 bps, as well as one low-speed data terminal transporting information at 16,000 bps, which in actuality operates as a 9600- or 14,400-bps packet transport. The remaining bits are available for specialized data signaling between customer-provided equipment and the network.
- Two telephones plus the low-speed data packet/signaling-type transport.
- Two dataphones, each operating at 64,000 bps, plus the low-speed data packet/signaling transport interface.
- One very high speed dataphone operating at 128,000 bps, as well as the low-speed data packet/signaling transport port.
- Notes:
    —Note 1—The dataphones can be operating in the circuit-switched mode where the path is maintained throughout the network or in the shared-packet transfer mode. Hence there are two types of dataphone offerings: circuit switched or packet switched for each of the above options.
    —Note 2—The dataphone can be a terminal, computer, cluster controller, printer, workstation, sensor, feedback control system, database server, facsimile device, polling device, etc.
    —Note 3—There can be any form of data-image-video information transfer at digital rates. Hence, the data-handling terminal could be an imagephone or videophone device, where video pictures or graphic images are simply sent as streams of data that effectively paint each row of the picture (coloring the dots (pixels)), where each pixel is a 16- to 24-bit datagram. These pixels are points (dots) of the video picture having multiple degrees of gray or full-spectrum color, depending upon the number of bits transported.

All these interfaces are summed up by the generic ISDN term 2B + D, where the B interface operates at 64,000 bps and the D at 16,000 bps. Here 2B means two 64-Kbps voice or data channels, where the two data channels can also be combined into a 128-Kbps data stream. Again, the information channels can be circuit switched or packet switched. The packet is simply a means of packaging streams of data with a header (of

X bytes—a byte is represented by 8 bits), a body (of Y bytes), and a tail (of Z bytes). Different protocols change these parameters, offering varying lengths (X + Y + Z) of packetized messages. In the packet mode, numerous streams of data are enveloped and packaged together and sent over shared transport facilities as they are stored and forwarded or simply switched to differing destinations. By being in a packet mode, they can be easily accessed by the transporter for specialized error correction, detection, sequencing, and routing control mechanisms to guarantee uninterrupted, safe arrival. Conversely, circuit-switched data means the pathway from point A to point B is specifically dedicated to a particular data call, thereby eliminating transport blockages once the original path has been established.

The public data network, based on ISDN interfaces and hierarchy, will use standards developed by CCITT international working groups. Interfaces have been defined so that terminal adapters are situated on customer premises to interface non-ISDN terminals to the network-terminating units. For the United States, there are two types of network-terminating units. NT2, located on customer premises, facilitates the movement of information channels and signaling to NT1, the network-terminating unit, where 160,000-bits-per-second traffic is transported back to the central office.

The ISDN hierarchy is designed to consider all communications for establishing physical connections, packaging, transporting, and presenting information. The International Organization for Standardization (ISO) has been working since the late 1970s to produce standards (known as Open Systems Interconnection (OSI)) to promote the free internetworking of information systems. This OSI reference model is structured into seven layers, namely the physical, data link, network, transport, session, presentation, and application layers. It is a layered hierarchical structure of communication protocols. The first three layers provide the mechanical, electrical, functional, and procedural characteristics needed to establish, maintain, and release physical connections and control the data flow, setup, path, selection, establishment, clearing, and management of the call.

The fourth layer, the transport layer, provides transparent transfer of information between end systems. It serves as a bridge between the lower network establishment layers and the higher information-formatting and processing layers. The higher layers are concerned with the call information in terms of encryption, formatting, synchronization, privacy, authentication, conversions, addressing, and presentation. (To help understand the interrelationships between these layers, some have used the analogy of writing a letter and transporting it through the postal system and presenting it at its destination, showing each step or layer of data handling from putting the letter into an envelope and routing it through the postal system to receiving it at the appropriate destination address, opening it up, and reading it.)

Network parameters

The extent and degree of the data-delivery mechanisms are specifically established to ensure that the overall transport and throughput objectives are achievable. They come into play to limit blockage, set transport-error-rate targets, and establish availability goals for data movement during high-usage, busy hours. To accomplish this, data transport parameters have a specified operating range for the type of network desired. Explicit parameter goals are especially needed for fostering more and more customer usage. Hence, each type of the following data transfer considerations operate within a specified range of values:

- Transfer time.
- Format conversions.
- Speed conversions.
- Classes of user services.
- Categories of error rates.
- Types of network signaling.
- Network synchronization.
- Network interfaces.
- Data call processing.
- Call clear—down times.
- Call request time.
- Numbering plan.
- Usage recording/billing systems.
- Inquiry handling.
- Route selection.
- Overall grade of service.
- Overall quality of service.

Data network features

Here is a voluminous list of expanding capabilities, ensuring the efficient/effective throughput of data:

- Broadcasting.
- Delayed delivery.
- Packet interleaving.
- Byte interleaving.
- Code conversion CCITT codes.
- Polling/sensing.
- Inquiry facility.
- Three-attempt limit.
- Low error rate.
- Data collection service.
- High grade of service.

- Standard interfaces.
- Data tariffs.
- Access to leased facilities.
- Duplex facilities.
- Bit sequence independence.
- Short setup.
- Call back (automatic).
- Redirection of calls.
- Speed/format transforms.
- Multiple lines.
- Abbreviated address call.
- Packet switching.
- Retry by network.
- Store and forward.
- Short clear-down.
- Manual/automatic calling.
- Manual/automatic answering.
- Data service classes.
- Direct call (hot line).
- Stimulus/functional interfaces.
- Barred access.
- Remote terminal identification.
- Multiaddress call.

Office and home ISDN interface arrangements enable up to eight terminal devices to share the access port.

Data services—data applications

These data network features provided by redundant, survivable, robust switching systems enable the network provider to package offerings and appropriately map new services to new applications. For example, network providers are now able to offer the following:

Numbering   Every data user will have a unique data address. There can now be a specific dataphone directory, similar to the telephone directory, in which each data customer's originating or terminating device can have its own unique number or, as noted earlier, up to eight devices can share a number with inward dialing arrangements. Depending on how internal CPE arrangements are shared, structured, and constructed, clusters of users, configurations of remote control systems, data PBXs, LANs, etc., can have external public numbering identities on the public data network.

Closed user groups   Groups of users can communicate among themselves and access outside external devices. The closed user group feature blocks outside calls from participating in this closed arena unless designated as part of a closed user group.

Broadcasting   An information service node can exist above the data network which, when accessed, will send out the message to multiple parties, depending on the current status of a changeable broadcast list. Alternatively, the broadcast list can be delivered with the message, and the network dynamically regenerates the message to the multiple parties.

Priority   Overrides and privileged handling capabilities can be provided to ensure that selected calls have preferential treatment. This feature came from military networks where emergency conditions required that current operating systems be interrupted with emergency demands or whenever a ranking officer had an important announcement, or to enable specialized data-call handling during high-congestion traffic conditions.

Delayed delivery   As part of the store-and-forward movement of packet information, a particularly endearing feature is the easy ability to delay delivery of the submessages in order to properly sequence the packages' arrival in correct order and ensure verification of error-free transport. From this service also comes the ability to store messages in the event that a receiving terminal is unable to accept them, such as out of order, out of printer paper, etc. As a result of these endeavors, both circuit and packetized information can be stored for later delivery, similar to voice-mail operations.

Data caller ID   Data messages contain originating-data-message-generator identification. This information is provided to the receiver, thereby enabling the receiver to have selected calls forwarded to specific changeable locations; other calls can be barred from accessing this customer. Additionally, authorization for access can be verified as security mechanisms invoke audit trails that determine if the actual authorized device at the actual physical location is indeed attempting to enter one's database. In addition, further identifying information contained in the header with the ID information enables the information to be transferred around the network to overlaying service nodes that provide, for example, language translations, etc.

Polling/sensing   Specialized CPE-network interface programs enable alarm monitoring, incarceration monitoring, energy management, meter reading, weather sensing, fire sensing, and monitoring of voting machines and opinion machines, etc. Such information is sent to higher-level service nodes for collection and distribution.

Functional services—applications

As we look across the major industries (banking, manufacturing, insurance, health care) and across market sectors covering both large and small businesses, state and federal governments, individual and multifamily residences, etc., we see the need for more and more data transport to enable everyday tasks to be handled electronically. This is accomplished by enabling data inquiry/response, data collection, data distribution, and data presentation mechanisms, as data is stored, accessed, transferred, transformed, manipulated, processed, and presented. Thus we see the need for overlaying the network with specific functional services that can be provided by both network providers and information-service or enhanced-service providers. Thus there will be common packages of offerings for specific computer-based data-handling services. They will be offered across the industries as data-handling application packages; these will enable electronic mail, fast facsimile (group 4), point-of-sale transactions/verifications, remote job entry, electronic information interexchange, electronic file transfer, electronic funds transfer, computer time-sharing, integrated voice/data workstations, audio graphic conferences, office communication via home computers (work at home), high-speed data entry, low-speed data entry, server-client application programs, database access programs, graphics, video storage, X-ray storage retrieval, electronic real-estate listings, video real-estate storage, entertainment program access, educational program access, personal health history file access, criminal history file access, emergency access to poison control center databases, federal disease center database access, 911 historic location data, compressed video transfer, videophone, specialized voice-recognition programs that enable voice-to-text word processing, security access clearances, etc.

Narrowband public data network options

Data will therefore be transported in either the circuit or packet mode providing the following options:

1. One 64-Kbps interface to a circuit-switched data network.
2. Two 64-Kbps interfaces to a circuit-switched data network.
3. One 128-Kbps interface to a circuit-switched data network.
4. One 64-Kbps interface to a packet-switched data network.
5. Two 64-Kbps interfaces to a packet-switched data network.
6. One 128-Kbps interface to a packet-switched data network.
7. One 64-Kbps interface to a circuit-switched data network, and one 64-Kbps interface to a packet-switched data network.
8. See Note 3.

Note 1 Offerings number one and four provide an additional voice interface to the public voice network in parallel to the data offering.

Note 2 All offerings provide additional access to a low-speed 9.6K or 14.4-Kbps packet network with additional stimulus/functional signaling up to the full 16-Kbps data transfer channel capacity capability.

Note 3 Option number 8 is simply two voice network interfaces with the low-speed data channel indicated in note 2.

Service pricing—network deployment strategies

After many years and numerous attempts to establish a successful pricing strategy for ISDN deliverables, it has become resoundingly clear that we need to price the service, not the interface. Additionally, uniform service deployment strategies are essential. The customer must not pay a separate distance fee to remotely access offices that are equipped for the data services, where other customers closer to the equipped offices pay less. Full availability in the area of delivery is essential. Islands of offerings are not sufficient. We need to establish a family of basic data-transport services that are supported by an appropriate packaging of selected public data network features. This will make the public data transport offerings attractive to the private data user instead of the low-speed, error-prone dial-up voice-grade transport.

Besides these basic service packages, there will be several layers of value-added "data-handling" services that can be provided via overlaying transport-changeable database mechanisms such as those offered by advanced intelligent network (AIN) vehicles or via separate, layered service nodes called application service centers. The AIN offerings will interrupt the data-switching transfers where additional routing translations occur via changeable databases. Both telephone company and third-party programs will coexist. Limits on these types of offerings are needed to ensure that the movement of data traffic is not inappropriately inhibited. On the other hand, the data calls can be switched via Open Network Architecture (ONA) tariffed interfaces to specialized information service providers (ISPs) and enhanced service providers (ESPs). They can perform additional work on the data call, such as specialized database lookups or adding unique data call handling capabilities.

It is essential in the movement of high-speed data that the network transfer be sensitive to congestion and delay so that the correct set of "in-line" basic services can be differentiated from additional data transfer offerings. Advanced intelligent network database updates and translation changes must not interfere with data flow. Customer choice/selection for long-distance data carrier, specialized routing for extended data transport services, or specialized database manipulations via layered service nodes need to be facilitated by some form of Information Network Architecture (INA) hand-off mechanisms that enable data calls to be routed to additional service providers, who process the information and then return it to the network for further transport without holding up the basic data call network operations. This ensures that any improper ESP/ISP system operations cannot cause any network degradation. Protective ONA interfaces between AIN providers must also be in place to appropriately protect the network from these call-flow interrupt types of service-node providers.

Choice of data transport across local exchange carriers' local access and transport areas (LATAs) must be available. However, these cross-LATA carriers must ensure that their inter-LATA/interstate message pricing does not improperly impose extensive data-call-handling costs that will drive the data customers away from using the public switched-data network. Similarly, long-distance providers across the nation need to ensure that their data transfer rates facilitate growth in the movement of local and national public switched data.

Local exchanges need to be equipped for data transport services that enable seven high-speed options for publicly switched, dedicated (circuit) or shared (packet) data transport, and these services need to be priced appropriately. In actuality, these seven options will most likely be reduced to either a circuit mode or packet mode data-transport offering that, with the addition of the low-speed packet offering, will cover all the options.

Pricing strategies must be for voluminous traffic, thereby encouraging the masses to generate massive amounts of data movement. This does not mean that the initial few data customers must support all delivery costs until the masses use the service (for example, four-year payback supported by a few customers). Providers must price for growth because it has been proven time and again that data users are very sensitive to price. This might require feature/service revenue-sharing relationships between network providers and equipment suppliers in order to quickly obtain the full range of needed services and attract the mass users. (For example, early ISDN pricing strategies required extensive startup costs to be passed on to the few initial customers, who were then offered only a few data-handling services by the initial data equipment suppliers. This type of offering was destined for failure.)

A public data network must offer a family of new services. It is in the best interest of the telephone companies and the national voice network users to remove data customers from the voice network, which was not designed for the different type of usage generated by data traffic. We must make the new data-handling options as economical as possible to encourage customer shift and rapid usage growth. Hence, high insulation costs, distance costs, limited availability, limited features, inability to access long-distance carriers, high usage fees, limited protocol interfaces, nontailored data transfer features, few service nodes, and nonubiquitous offerings within a local area must change. Network deployment and support capabilities, as well as competitive pricing strategies, must be part of the delivery considerations for successfully launching a new parallel, narrowband, public, data-switched network with both circuit- and packet-mode data transport. This might result in a multitier usage-pricing strategy in which the small business and occasional home user are provided with an areawide/statewide basic-rate service, while small and large business users, generating greater usage, are offered higher usage blocks of economical data transport. It is equally important that all offerings ensure access to advanced/enhanced data services by AIN or INA interfaces. Otherwise, users will be driven to obtain these services from internal CPE devices (for example, broadcast, where the CPE device simply pumps out numerous messages via an internal list of external data users).

So as providers encourage prospective customers to shift from their current modes of operation to exchange fully switched electronic information, providers must do everything possible to promote rapid acceptance and growth. Massive advertising will be required to not only tell the prospective customers that a public data network is now available, but also to show the customers how to use it and how it can assist them in their daily tasks.

Provisioning and support centers from the network provider are needed to allow the data-customer service-order representatives to appropriately identify needs and quickly obtain a matching service package. Customers also need to be assured of service availability. So while discussing the various data-handling packages with the customer, the service representatives should be able to determine if the transport facilities to the customer are of the data-handling quality. This might require a quick, automated test of the facilities to ensure that timely provisioning is possible. Once customer acceptance is obtained, data service orders need to be automated, thereby ensuring rapid service delivery.

Ubiquitous service availability is key. Other countries have shown that it is essential to provide the service throughout the country. LECs, both RBOCs and independents (such as GTE), are now positioning themselves to establish the public data network. LECs must work together and with other common carriers (IXCs), as well as interested value-added network providers (VANs), to ensure that sufficient service access and interconnection mechanisms are in place to enable a customer in one region to easily interconnect with another in another region. Otherwise, the services should not be prematurely launched, leaving the customer in a quandary. This could be a key role for Bellcore and the RBOCs' and other independents' standards groups to facilitate.

Additional considerations

Information arena—layered networks' layered services When developing the public switched-data-transfer information arena, it must be constructed in a manner to facilitate a layering of networks that provide a layering of services. For example, a doctor within a medical building using an internal private local area network or data-switching information branch exchange (IBX) wishes to access the local public data network to route the doctor's query to a medical service node. The service node in turn sends several message inquiries to individual hospital service centers, which search through their specialized databases for the particular patient information desired by the doctor. The overlaying medical service node uses the network to access more specialized CPE service nodes within the hospitals; each layer of service further enables the completion of the data search. There are thousands and thousands of such applications in which data calls use various CPE, LEC, VANs, IXCs, ISPs, and ESPs' systems to obtain the desired solution by obtaining layers of services offered by layers of networks.

Centrex/PBX-type customer data-handling arrangements Numbering plans for a public, switched-data network are essential as the world shifts to directly switched, addressable ISDN-based public data handling. Even local area networks are being interconnected via narrowband ISDN arrangements. We will see more and more private-to-public-to-private switched data transfer arrangements grow substantially over the latter half of the 1990s, especially as fully switched offerings become economically available to interconnect cities, states, regions, and the nation. In this expanding arena, fully switched data PBXs and data-handling Centrex systems will enable both voice and data calls over the same twisted-pair internal wiring facilities, which will not require separate coax wiring.

Wireless data It is quite evident that some data customers require both fixed-location and mobile-access data-handling arrangements, especially the building contractors and tradespeople, who many times wish to order material while visiting or working at the remote work site. There, the truck or van becomes the mobile business office. Today, much of this work is performed verbally standing at grocery store pay-phones, where customers wait in line for access to basic telephone service. In the future, these operations will shift to verbal cellular calls with the expenses passed on to the home or office buyer. The spectrum might be tied up for long periods while waiting for distant operators to access and change internal databases. By providing direct access for the remote tradesperson or contractor at wireless 64,000-bps rates, ISDN data calls can be economically justified to speed up transactions, ensure security, remove errors, dynamically facilitate better choices, etc. This type of integrated voice-and-data mobile offering will be a viable solution to many business users, large or small.

Voice-grade data Indeed, there is a need for transporting subrate data to and from customers who do not have ISDN-compatible terminals or where ISDN interfaces are not yet deployed and voice-grade modems are required. To facilitate both endeavors, provisions have been made to interface and transport such asynchronous data, but care should be taken to limit this subrate traffic from deteriorating the overall grade of service for both the data and the voice networks. For example, low-speed CPE data does not assume an error-free network. Very small blocks of information are transported at low transfer rates, requiring extensive acknowledgments and repeat requests. This type of movement slows down the 64,000-bps data transport mechanisms' effective throughput rate and also causes severe "brownouts" and noise conditions imposed on the movement of voice-grade traffic.

Global data Once the local data networks are in place, access to data-handling value-added networks, both nationally and internationally, is mandatory. Gateway services will be layered on local transport mechanisms, providing access to global transport service nodes that assist in the addressing, routing, and secure delivery of the data calls. Additional global application service centers will enable specific global community-of-interest services for the various market sectors, such as global medical service centers, enabling specialized access to remote databases for authorized queries. Hence, there will indeed be a layering of global networks, providing a layering of services as many players touch a data call and provide their own unique twist or turn on it to ensure successful completion of its journey. To accomplish this, it is important to establish public data network gateways to the many differing global networking interfaces.

Private-to-public networking Today, due to an absence of data-handling PBXs and a data-handling public network, there is a proliferation of LANs interconnected via leased-line/trunk transport. As bandwidth becomes more economical, there is a trend among common carriers to compete by offering usage-rate, shared transport. The future will see voice-grade PBXs give way to higher- and higher-speed switched-data transport platforms that are integrated with voice switching. Already, PBXs are being equipped with ISDN interfaces and internal data-switched matrixes. Frame relay and SMDS shared-transport offerings are being supplemented by asynchronous transfer mode (ATM) and synchronous transfer mode (STM), fully switched data offerings. As the internal customer networks change over to fully switched broadband services, future PBXs will have ATM/STM platforms. These technical possibilities are discussed in detail later on in this analysis. For the moment, suffice it to say that fully switched narrowband data-handling capabilities using the same twisted pair for both voice and data and interfacing to a fully switched public data network transport vehicle is an attractive alternative for private-to-public data networking.

In conclusion

In this short analysis, we have not addressed all the questions pertaining to data transport. For this, we must turn to the reference works to obtain a full, clear understanding of the networks, the services, the opportunities, the issues, and the strategies. We have, however, attempted to provide a quick overview, noting in general terms: What is data? How is it different from digitized voice? What is circuit data versus packet data? What does ISDN offer? What are its data-handling options? What is a public data network? What are its parameters? What are its data-handling features? What specific services can be obtained from these features? What about deployment strategies and pricing issues? What are some specific data transport service offerings? What services can be obtained from overlay service nodes such as application service centers? What are the roles of wireless data, private-to-public networking, data PBXs, and global data networks?

A public digital voice network

As indicated earlier, we have seen how integrated digital networks (IDN) convert voice conversations into digitized 64,000-bit streams. Adding the data services to IDN establishes ISDN, but it is important to note that ISDN provides not only new data, but also new voice transport capabilities. ISDN arrangements enable a single twisted pair to transport voice conversations in specially coded digital pulses at transport rates of 160,000 bps. At this rate, we can obtain two voice conversations (2×64K) as well as a 16,000-bps low-speed data transport channel, where additional services such as voice Caller ID services, can be easily facilitated. (For example, depending on the Caller ID received, the called party might request specialized handling of the digitized voice call, such as having it processed through language translators or having selected calls sent on to another destination such as a car phone or personal phone.)

Once the voice call is digitized from end to end, this greatly improves the quality of the call in this relatively error-free environment, eliminating many of the traditional telephony analog delivery noise problems. Thus, besides being able to have a second line to the home, over the same twisted pair, we are able to deliver 7-kHz stereo-quality audio signals for digitized musical entertainment. There will be numerous additional advances in the transport of voice calls, whereby voice-storage, voice-to-text, secure-voice, voice-call-blockage, obscene-phone-call reroute/tracing, and voice-broadcast services become easier to achieve and more economical to deliver via the narrowband digital voice information transport network. Finally, the ability to talk and simultaneously look at data provided by integrated voice-and-data delivery mechanisms offer exciting changes to our everyday mode of operation. There will be equally exciting economical narrowband videophones, as well as full multimedia voice/data/video workstations. Indeed, the narrowband information network is the first step, a major step, to an exciting new information society. It is time to leverage our existing copper-based delivery system to quickly enter this new age, the information age.

3

Wideband information

"Of the wide world . . .
dreaming on things to come."
William Shakespeare

Better, faster, cheaper, error-free, ubiquitous public data transport begets more and more universal data access, storage, manipulation, and presentation. This, in turn, begets more and more data transport, which begets more and more data access. As this occurs, customers will seek to interconnect local area networks (LANs), wide area networks (WANs), metropolitan area networks (MANs), remote workstations, central computer mainframes, physically distributed databases, and supercomputers to rapidly exchange information. As videophone conversations require less compression and greater resolution, as more and more narrowband information needs to be exchanged, as complex technical solutions address every aspect of the marketplace, we need to meet these ever-increasing telecommunication expectations with greater information-handling capabilities from our existing and future plant. This expansion can initially take the form of an overlay network, but it must be the predecessor to the fully switched broadband information network. Hence, wideband's transport topology and switching hierarchy must provide the building blocks upon which the final broadband structure will be constructed. To accomplish this, many questions need to be addressed.

What are wideband information networks? Isn't private LAN-to-LAN networking simply a leased-line/trunk, port-to-port offering? What are T1 or T3 networking options via digital cross connects and via SONET? What are fractional T1 and fractional T3 offerings? Explain virtual private networks. What are the differences between permanent virtual private networks and switched virtual private networks? How can the network providers enable

the customers to dynamically change bandwidth and destination requirements? How can they reduce setup times from months to weeks to days to
hours? How do we achieve multimedia combination: voice, data, and/or
video messages? What is the role of frame relay—locally and nationally?
How does this shared network offering meet private needs? How does
SMDS, a shared-usage transport vehicle, integrate with various switching
facilities to achieve a trans-switching offering? (See appendix C.)

What are the roles of FDDI, Ethernet, ATM, and STM? What are the
new ring switches, edge switches, and class 6 switches? What services do
they offer? How do we obtain private-to-public-to-private internetworking?
How do we achieve a fully switched public wideband information network
(PWIN) locally (urban, rural, state), regionally, nationally, and internationally? How do the LECs internetwork with VANs, LANs, IXCs, ATPs, CAPs,
and SBSs? How do we ubiquitously deploy the public wideband information network? How does it interface to customer-premise networks such as
the new ATM PBXs or class 7s? How does this PWIN meet the needs of
state governments; the medical, legal, and other communities of interest
(COIs); and the large and small business sectors? What is the plan of action
for obtaining a fully switched, addressable offering—urban, rural, national,
and global? How does PWIN form the supporting structure for future
broadband switched services over fiber-based SONET facilities? With what
new topology? Requiring what new fiber-deployment plans?

Background

So how did the telecommunications industry arrive at its present situation
and what is required to move it forward? First let's take a backward look.
By not providing a fully switched, data-handling, internal PBX-type network within customer premises, data customers were forced to manually
(hand carry) IBM cards to their internal, large mainframe computers as
they initially ran their programs in the batch-processing mode of operation.
Soon various "online" input-output mechanisms allowed several nearby
terminals to input separate programs, which were placed in separate fixed
or variable-sized partitions of the computer and executed in a shared-pro-
cessing manner (as a prelude to parallel processing). As this technology
became more sophisticated, internal bus structures within the computer
were extended further and further out to remote locations. This enabled
numerous terminals within the office complex to interface directly to the
computer and exchange information via the computer's ever-expanding
shared databases in "real time." With these new data telecommunication
capabilities, the local area network (LAN) was born.

During the late 1970s and 1980s, there was rampant development of local area networks as the computer was integrated into the office complex.
Every conceivable list of information was electronically stored on tape,

drum, or disk for easy access and search. Computer sales blossomed as computer power doubled every 18 months. Processing changed from singular in/out methods to simultaneous foreground/background input/output techniques, and then to more sophisticated, physically distributed multiprocessor arrangements. To facilitate this shift, remote terminals were no longer simple keyboards and printers. Remote terminals became sophisticated personal computers, performing extensive data manipulation and presentation and enabling people of differing skills to participate in the solution of problems. To accommodate this, LANs needed to exchange information throughout the office complex by bridging from one to another and then by using sophisticated routers to send information from one location to another within a local building complex or campus. To cross town, LANs used leased lines. Otherwise, they used local or foreign exchange (FX) facilities to dial directly. These low-speed data modems (modulators/demodulators) transported information at 300 characters a second (2400 bits per second) or at best 9600 bits per second to send 1200 characters a second (approximately 200 words per second, a far cry from 100-wpm teletype telex messages). This was the world of wide area network arrangements, where data was transported across town or across the country, via the public voice networks using voice-grade data-handling methods. Over time, specially conditioned facilities enabled clusters of users to send information directly across the country at 14,400 bits per second or 19.1K (K = 1000) bits per second. Finally, 56,000 (56K) bits per second became a specialized, high-priced analog, high-speed data-transport service.

However, pressure mounted for higher- and higher-speed facilities. So T1 was developed, which provided digital, long-line transport capabilities for moving 24 digitized voice conversations, each at 64,000 bits per second. This also became the tool for data transport. T1 could handle a digital data stream at 1.54 million bits per second (actually 1.35 million (with overhead) bits per second). After T1 came T2 at 6.3 million bits per second. Both were transported over copper-based facilities. Next came T3, which required fiber to move information at 45 million bits per second. Using these leased facilities, data customers could "pump" their data streams down these digital "pipes" at relatively error-free rates from point A to point B.

Unfortunately, it took considerable time to set up these facilities and provide the desired conditioning to the users' locations. Sometimes it took as long as 30 to 60 days to properly achieve access to all the national hookups. These transports also became quite expensive. For example, a T3 link from Minneapolis to Denver could cost as much as $75,000 per month. To reduce costs, various compression techniques were deployed to reduce transport requirements. However, pressure again mounted to be able to quickly change configurations such as from location A to location C, A to B, or B to C. This then led to shared private networking solutions where specialized switching systems were located at centrally located, internal customer premises in order to separately direct private traffic from location A to either location B or location C, as well as between locations B and C. Over time, as internal operations attempted to share more and more leased transport, pressure mounted and mounted.

Present status

There were pressures for more and more internal control, and there was an ever-increasing appetite for more cost-cutting transport expenditures. From this came the need to use only a selected amount of T1 or T3 facilities for a given period of time (for example, using only 384 Kpbs, 640 Kbps, or 784 Kbps of T1's 1544-Kbps capacity, or using only two of the possible four T1 capabilities of the T3 facility). These options are called fractional T1 or fractional T3, even though the facilities that are provided directly to the customers are capable of the full T1 or T3 data movement. This is similar to using only 56K of the 1.5-Mbps frame relay access facility. Hence the desire exists to send a variable number of channels, each carrying 64,000 bits per second. (See appendix C.)

Digital cross connects were used to automate the provisioning of these transport paths, as telephone office mainframes, which originally manually interconnected paths from the customer to either direct or switched facilities, were automated. Digital transport systems multiplexed digital channels together that could be easily assigned to one or more customers. Using remote control capabilities, customer traffic was combined at remote gathering points (nodes), located closer to customer locations, in order to share transport facilities. Using these same switching arrangements, data paths could be quickly set up, thereby reducing service order provisioning time to days, and in some cases hours. To help facilitate and advance this operation, a new digital technical capability was added to transport voice and/or data together in the form of 23 separate channels with a signaling channel, each operating at 64,000 bits per second. This type of service was called primary rate ISDN or 23B + D, where the B means information handling at 64,000 bits per second, and the D means signaling information handled at 64,000 bits per second. Sometimes this is referred to as wideband ISDN (W-ISDN). In this manner, a PBX on customer premises can combine voice and data traffic into these separate B channels, with the signaling channel—the D channel—telling the network exactly what information is on what B channel. Unfortunately, in actual implementation, the voice calls are not separately removed from this transport system and switched onward upon receipt by the public voice network. Presently they remain on the private-network leased facilities. This was not the original intent. (We will see how W-ISDN should become an interface to a fully switched wideband network.)

Another type of service, called switched multimegabit digital service (SMDS), was defined by Bellcore in the late 1980s. The service was based on the technology developed from metropolitan area networks (MANs), where information from multiple customers (LANs) could be transported over shared facilities at rates of 4 megabits, 10, 16, 25, 34, . . . 45 million bits per second. Later, this transport will enable new SONET rates, which are basically multiples of 50+ million bits per second. Another technique, called *frame relay*, is specifically targeted for moving information from one LAN to another at rates of 56K/64K bits per second to 1.5 million bits per second. This accommodates most packages of varying frames of information from the various LANs. Frame relay users are offered expanded bandwidth capabilities that might or might not be available during congested, busy-hour times. This could cause droppage and loss of various packages of information. However, this sharing of transport does reduce long-distance transport costs, thereby achieving a circuit from point A to point B of somewhat virtual bandwidth. Though this is not a physically held path, it effectively becomes a permanent or dedicated path of dynamically changing bandwidth over shared facilities from point A to point B.

Future endeavors

Switched frame relay circuits can be deployed to switch the calls from location A to location B, C, or D dynamically over shared transport. Similarly, using SMDS protocols, data can be switched to variable destinations by what are called fast packet switches or asynchronous transfer mode (ATM) switches. Both SMDS transport and ATM switches move packets of information in packages of 53 bytes. (Remember a byte contains 8 bits of information. The header contains 5 bytes and the body contains 47 bytes of information.) On the other hand, synchronous transfer mode (STM) switches enable circuit-type paths (similar to today's voice calls), allowing information to be transported at 64,000 bits per second, 1.5 Mbps, 6.3 Mbps, 155 Mbps, or 620 Mbps. In time, these switching platforms will enable variable numbers of 64-Kbps channels to be sent, thereby covering this entire spectrum.

Thus, a videophone conversation, which consists mainly of a talking head, might only require 1.5 million bits of transport or even 128-Kbps or 384-Kbps facilities in the compressed form, depending on screen size. However, as the head moves, this will require more bits that might necessitate 45 million bits of information transfer. Thus, it consumes more or less bits as needed during the conversation. Similarly, bandwidth requirements will change as one supercomputer talks to another and sends, for example, the design of the hood of a new sports car from a remote design location to a centralized supercomputer to see if it fits properly into the new structure. This dynamically changing information can be readily accommodated by the channel switches that dynamically change the direction of transport from location A to location C or to location D for varying amounts of information. (See Fig. 3-1.)

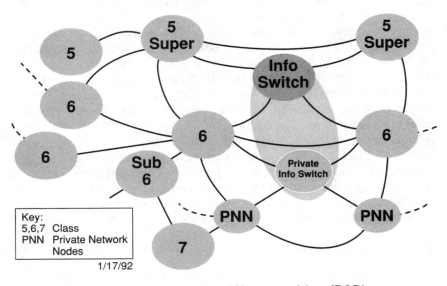

Fig. 3-1. Private and public networking (P&P).

So there are two arenas for this dynamically changing data. The initial arena, the wideband, enables the customer to instantaneously establish paths to various destinations and send variable amounts of information at rates of "n" number of 64,000 bits per second up to 45 million bits per second. This pre-SONET (synchronous optical network) environment is called wideband. Subrate SONET would use virtual tributory payloads for less than DS3 rates. Once we have all the transport standards in place for SONET-type transport at multiples of 50 million bits per second, we then enter the broadband arena, where B-ISDN interfaces and switching hierarchies are in place. This will be augmented with the broadcast broadband cable facilities. Thus, fully switched, fully interactive, fully addressable wideband facilities are needed as the stepping stone to future broadband endeavors. (See appendix B for a more detailed technical analysis.)

Assessment

Many possibilities exist for rapid movement of business data from one location to another and the transfer of high-speed, better-resolution videophone calls. So what type of network should be universally, publicly deployed that will accommodate more and more wideband/broadband transport? What are the next steps to achieve it? As fully operative fiber is deployed, enabling megabit transport, what topology and switching hierarchy should be deployed to ensure that wideband is indeed the correct building block for broadband?

Wideband information users

Who are the key users of the future fully switched wideband information network? They are the large business community, the education community, the state governments, the federal agencies, and the occasional sophisticated work-at-home or small business person. Most will require fully switched 1.5-million-bit-per-second capabilities that transport considerable data to and from workstations, LANs, supercomputers, databases, video files, etc. over the existing twisted-pair, specially conditioned copper-wire plant.

As more and more compression algorithms make it possible to send better-resolution video at 1.5-million-bits-per-second rates to enable television-screen-size capabilities, there will be a continuing extension of these offerings into the residential areas. But care should be taken to differentiate narrowband from wideband and broadband applications. While the narrowband dataphone and videophone services have a strong role to play in the small-business marketplace, wideband dataphone, videophone, and videoconferences have a key application in the large-business and government markets. On the other hand, the residential communities are prime candidates for both narrowband dataphone and videophone, as well as selected wideband dataphone services. Though the existing plant to the homes might be upgradable to handle wideband videophone, many believe this step is only an interim step on the way to fully interactive, broadband, high-quality, high-resolution offerings that will be coupled with high-definition television broadband services, using fully interactive B-ISDN 155/620-million-bits-per-second facilities.

This does not mean to imply that businesses will not also want broadband videophone when available. In general, each community of interest can economically meet its internal needs by using variations of existing wideband technologies as they wait for the fully developed, robust, survivable broadband switching and service platforms that will be discussed in detail in the next section. Hence there is a time and a place for wideband as we pursue the most economical use of existing plant while designing and deploying the forthcoming broadband facilities. By doing this, we will address many of the existing needs as we prepare for future needs of the marketplace.

Wideband information user needs

Wideband customers are substantially different from narrowband data users. They are no longer content with moving small amounts of data to construct messages or search databases for singular names or records. They are moving higher forms of intelligence, specifically in the image, graphic, and video form, as well as large databases and program information among supercomputers and workstations. Thus, to emphasize these higher transport needs, these systems are called wideband information networks as a prelude to the broadband information networks, different from the narrowband information networks, discussed earlier. These wideband users need to:

- Transport information from narrowband rates to T3 rates, at multiples of 64,000 bits per second—from 128 Kbps to 45 Mbps. As noted, the T1, T2, and T3 standards are simply a prelude to the Synchronous Optical Network (American)/Synchronous Digital Hierarchy (European) (SONET/SDH) standards. In time, these wideband rates will simply become the subrates of the SONET hierarchy, which will be multiples of 50 megabits per second.
- Dynamically require variable bandwidth (more or less) initially at call setup time.
- Establish permanent virtual circuits (PVCs). This means to be assured of a path from one location to another during congestion periods as a dedicated or permanent hookup over a shared facility, transporting a predefined maximum amount of information.
- Be able to dynamically change outgoing destination from location A to location B or C, or whatever. These are called switched virtual circuits (SVCs).
- Need to dynamically change the amount of bandwidth required during the call so the customer has a virtual bandwidth capability, even though the full amount of bandwidth required is not specifically held at all times for the individual customers.
- Dynamically change destination locations at the beginning of the conversation, as well as dynamically change bandwidth during the conversation.
- Have survivable transport that can be handled by different switching centers in the event of fire or destruction.
- Be able to access switching nodes to obtain shared transport at locations closer to the customer and remove unshared distance costs to the central office facilities.
- Have access to alternative carriers via the public network from locations as close to the customer as possible, thereby reducing the last-mile bottleneck.

- Have transport that is insensitive to blockage and droppage of frames of information, thereby requiring a high grade of service from network providers to ensure that relatively error-free through-put is achieved.
- Have interfaces to local switching nodes from differing local area network protocols.
- Have the ability to differentiate and separate wideband ISDN voice calls from data calls in order to move the voice calls to the public digital voice network and data calls to the public narrowband data network or the public wideband information network.
- Have identifiable, separate wideband-user addressing capabilities similar to narrowband addressing structure.
- Have wideband information user directories, both electronic and yellow page varieties.
- Have network-to-network access mechanisms for LEC interface to IECs and VANs.
- Have service node access mechanisms to obtain specialized wide-band information-handling services and access to specialized im-age, graphic, and video information bases such as education on demand, X-ray files, travel logs, and sporting events.
- Have multimedia services between CPE and public network, CPE and service nodes, CPE and alternative providers (VANs), CPE and CPE for integrated voice, data, text, image, and video offerings.
- Have broadcast capabilities for incoming "live" or delayed-delivery video services.
- Ensure privacy.
- Have secure and survivable (S&S) transport.
- Have universal access and service availability throughout the city, state, region, nation, and globe for the wideband information network.

Proposed solution

To accommodate these wideband user needs in terms of using, modifying, and expanding existing transport capabilities augmented with fiber, and to prepare the proper building blocks for supporting the future broadband platforms, we need to:

- Overlay a digital wideband information transport switching hierarchy over the traditional analog telephone company network, as the net-work is being initially changed to a narrowband IDN-ISDN network.
- Establish new access switching nodes closer to the user for provid-ing the first point of entry to the public wideband information net-work. This can be referenced as the new Class 6 offices, since they are survivable and can route traffic to multiple destinations.

- Construct these survivable access switching nodes on survivable rings in hardened sites (huts protected for fire, etc.) or as collocated switches in central offices, thereby achieving direct interconnection among remote access switching nodes.
- Enable wideband information traffic to be routed directly from these access nodes to an IXC, VAN, ATP, or CAP. Also, wideband traffic can be routed from these access nodes through higher-level, class-5-type offices with collocated wideband information-switching capabilities to interface to these networks. Remember that the traditional voice network switching hierarchy is such that a customer-premises phone interfaces directly to a class 5 office, which in turn connects directly to another class 5 or a local access tandem to route the call directly to an interexchange carrier class 4 office. That office might then route the call to another 4 or to a class 3, 2, or 1 to send calls coast to coast and then back down the hierarchy to the destination phone.
- Enable the following switching options:
  —CPE—Class 6—CPE
  —CPE—Class 6—Class 6—CPE
  —CPE—Class 6—Class 6—Class 5—Class 6—CPE
  —CPE—Class 6—Class 6—Class 6—Class 5—Class 6—CPE
  —CPE—Class 6—Service Nodes (ASC)—Class 6—CPE
  —CPE—Class 6—Class 5—ASC—Class 5—Class 6—CPE
  —CPE—Class 6—IXC/VAN/ATP/CAP
  —CPE—Class 6—Class 6—IXC/VAN/ATP/CAP (see appendix B for acronyms)
- Enable a direct interface from CPE information PBXs (that will use ATM-type switching systems) to the remote-access switching nodes.
- Enable direct interface to CPE local area networks (LANs).
- Enable CPE primary rate ISDN transport mechanisms with integrated voice and data messages to interface to wideband access nodes for further switching and separation of voice/data traffic.
- Enable signaling channels to transport call service content information so differing providers can offer value-added services to the call. This enables the call to be dynamically directed by initial call setup procedures.
- Enable security, survivability, privacy, authentication, validation, and tracking-type mechanisms to protect and ensure the proper movement of the information.
- Enable service nodes and other access providers to reach the customers at the lowest point of entry into the network, thereby reducing the local bottleneck as much as possible via appropriate transport, addressing, and billing arrangements.

- Price for growth of wideband information movement, while ensuring that congestion, blockage, and information losses are kept to the minimum.
- Enable access to HDSL/ADSL-type local loop arrangements to enable local plant to specifically handle 1.5-megabit information traffic.
- Ensure universal availability throughout the city, state, and region with minimum distance access penalty.
- Enable interfacing with private networking transport protocols, such as frame relay, SMDS, FDDI, and internal CPE information-based PBX switching centers.

Competitive arena

Wideband private networks will continue to expand as dial-up wide area networks (WANs) give way to these fully switched variable-channel megabit possibilities. There will be several types of participants. One form will be on customer premises consisting of a privately owned nodal switch. The traditional voice PBX shifts to an information voice/data/video private branch exchange (I-PBX/IBX), which most likely will exist in a fully distributed form with separate modules located throughout the customer premises. Its modules will be located on separate floors or in separate buildings throughout the local campus. This new I-PBX/IBX will require universal addressing and internal addressing. It might, in time, be referred to as the new class 7 as part of the need for more and more internal facilities to be directly addressable and accessible in the global information marketplace. These internal systems will obtain public access via the new public switching nodes located closer to the customer premises (earlier referred to as the new class 6 switching nodes). See Fig. 3-2.

Alternative private switching nodes might be established on centralized customer facilities that are used to collect information traffic from various area-wide customer locations to interswitch these facilities and provide direct access to the world, interdependent with resident local exchange carrier (LEC). These hubs might be part of the customer-premise equipment (CPE) or provided by alternative transport providers (ATPs) or competitive access providers (CAPs). In time, shared gigabit rings might develop throughout the local community. These will be partly private and partly public. In any event, as the FCC encourages more competition within the local arena, it will indeed be a competitive arena, but care must be given to ensure that not just the most lucrative areas foster information competition. Other areas, such as rural communities, should be provided wideband information networks as well.

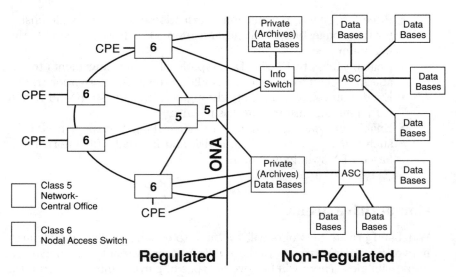

*Fig. 3-2. Future regulated/nonregulated architectural boundaries con-
siderations.*

In conclusion

The public wideband information network (PWIN) will be overlaid on the
existing plant, requiring 10%–20% of the present facilities to be appropriately conditioned to transport information up to rates of 1.5 million (T1)
and perhaps 6.3 million bits per second (T2). Fiber can then be selectively
deployed to enable the initial T3 (45 million bits per second) transport
rates and eventually be the basis for a SONET broadband transport network. A new switching platform will be overlaid to extend the existing five-
level hierarchy to include a new sixth-level switching node located closer
to the customer and operating on a survivable ring-type structure. These
nodes can be interswitched at this level or subsequently routed up through
the hierarchy via the fifth level. The nodes can also be directly interfaced
to the interexchange carrier (IXC), VAN, CAP, ALT, a specialized service
node (application service center), or a local CPE (LAN or I-PBX) network.

Information will be dynamically routed and switched in multiples of
64,000 bits per second up to 45 million bits per second. These wideband
rates will eventually be subrates of the broadband SONET optical carrier
(OC), where OC-1 supports 50+ million bits per second and OC-3 supports
155 million bits per second, etc. Customers will be offered permanent virtual circuits (PVCs) and switched virtual circuits (SVCs) to initially enable
universally switched LAN-to-LAN traffic via Ethernet, frame relay, SMDS,
and FDDI transport protocols. However, as fully interactive multirate ATM
and STM switching capabilities become available, wideband information

users will have global addresses so they can send and receive variable-rate wideband information calls. These calls can be processed by enhanced service nodes (application service centers), offering particular specialized call handling. All in all, a new, fully switched public information network at wideband rates up to 45 million bits per second could be established across the country using upgraded, specially conditioned existing plant, with additional fiber as needed. This offering will be a subset of the fully switched broadband ISDN (B-ISDN) and SONET/SDH-based public broadband information network.

4

Broadband information

"This indeed . . .
Is a singular broad opportunity . . .
Missed by many an untrained eye . . .
Let us proceed with haste . . .
The game is afoot . . ."
Sir Arthur Conan Doyle

During the many years since divestiture, much has been said concerning broadband opportunities. Much has been technically accomplished to prepare for a new information-handling infrastructure for a new age—the information age. Now is the time to step back and reassess the situation. It is time to bring the full range of technical possibilities and market opportunities into proper perspective. A limited view, a view that is focused only on one area, or a view that is slightly out of focus could cause a subtle shift in emphasis in execution. This could have multiplying, devastating results on the ultimate technical infrastructure and on the competitiveness and the availability of future service offerings. These services will be needed for the public and private arena, the local and global marketplace, the network and customer-premise equipment suppliers, the residential and business offerings, the state and federal governments, and the service provider and the provider's customers today and in the future. Now is the time to differentiate between present and future coax and fiber capabilities, wideband and broadband technologies, and broadcast and interactive broadband services. So as we begin our formal pursuit of a public broadband information network, let us first consider where we are going and where do we want to end up—with what network infrastructure? Supporting what services?

We need to address the following key questions: What is a public broadband information network? What services will it deploy? For what applications? Using what technology? Asynchronous transfer mode (ATM)? Synchronous transfer mode (STM)? What is the difference between broadcast and interactive broadband services? How will the broadband network be deployed, automated, overlaid, and integrated with the current voice-grade network, the future narrowband and wideband networks, and the existing cable network? What is the future fiber deployment topology? How will broadband calls be handled? With what type of new switches? How will the services be secure and survivable? How will we use the forthcoming technological advances, such as 100 wavelengths per fiber? Where will the network be deployed? Over what time frame will it be phased? How will it be supported? How will CPE terminals, LAN transporters, information PBXs, and customer-controlled switching nodes (class 7s) access the network? What will it cost? How will it be achieved (funded)? What revenues will be obtained for the following services: videophone (broadband), high-definition television (HDTV), multimedia workstations, computer-aided design, computer-aided manufacturing, computer-aided engineering, supercomputer networking, image/graphic design transport systems, global networking, and better facilities to handle high-volume narrowband/wideband traffic? What specialized networks for research/education facilities will be overlaid on the public superhighway? What plan of action should be pursued to achieve what services? When? Via what service nodes? Application service centers (ASCs)? For what communities of interest (COIs)? What access gateways are needed for what global networks? For what database accesses? What role will alternate transport providers and competitive access providers (ATPs/CAPs) play with the various types of private-to-public networking arrangements? What service node switching arrangements are required (info switches)? What broadband network capabilities will need to be available in 2000, 2010, 2020? What are the new service objectives, goals, and key strategies? What is the plan to achieve them? When? Where? Why? For whom? How? With what RBOC organizational restructuring and cross industry alliances, partnerships, mergers, and acquisitions? What what specialized revenue-sharing plans? What regulatory relief? For what short-term and long-term revenue targets? What is the impact of a public broadband information network on society? The rural communities? The United States? The world? What's in it for shareholders, employees, management, and the customers—the new information users?

A public broadband information network

So let's begin at the beginning by analyzing the potential users—the customers and their broadband applications. Then we can review the technical possibilities. Let's first determine if there is indeed a need for a public broadband information network. Should it initially only have wideband capabilities? Should it simply be a broadcast network similar to present-day cable and satellite offerings, providing entertainment services with a few more choice offerings? Or should it be constructed using building blocks that formulate a fully interactive infrastructure that grows in capabilities as unlimited technical advances take place throughout the twenty-first century? So let's now assess what we need, what we want, and what our lives will be like in an information society with a public broadband information network.

Broadband applications

Wouldn't it be nice to have a high-quality video conversation with your mother, father, husband, wife, son, daughter, grandparents, boyfriend, girlfriend, boss, or co-worker? Wouldn't it be exciting if you could see each other with the same quality and resolution as being there face-to-face? No matter how fast there was body movement or how quickly the conversation flowed, the voice and mouth would be in complete synchronization, and there would be no blurriness or disruption of the images. Background lighting, skin tones, camera distance, and other considerations would be appropriately handled to provide a comfortable, nonrestricted conversation. When the person at the other end of the videophone call can successfully observe the caller's new dress, new pearls, or brooch, or when the fine signs of tiredness and fatigue can be clearly seen and understood, these items can help to determine the mood and receptiveness to discussing new ideas or solving old problems.

Do we want to live in a world of small-screen, black-and-white images, or full-size, wall-type, full-color images? Do we want to find out in the year 2010, after the entire network is completed, that no one can take advantage of the new customer-premises equipment (CPE) service that requires higher bandwidth availability because we had originally built the network to only pass a limited number of interactive, high-bandwidth services? Wouldn't it be nice to enter a round entertainment room in the home and feel that we are actually at Orchestra Hall, climbing Mount Everest, or traveling down the streets of San Francisco in a trolley car? What will it be like to stay at home on a cold night during a Minnesota snowstorm and remotely participate in a class and be able to look around to see who else is in the classroom and electronically sit next to them to have discussions at the break? How will it be to sit at an electronic conference table and have an electronic meeting with participants from around the world present in holographic form? Why should firms continue to add more and more floors to downtown office buildings and cause highly congested areas? Why not work in satellite locations ringing the city, using shared office space interconnected by broadband facilities? Why not return to living in rural environments and maintain employment within a virtual firm, one that does not require traditional work locations but uses communications to inter-network employees from their homes? Why should my children and my children's children continue to live in large cities with approximately 97% of the population, leaving only 3% in the rural areas? These cities were formed in the 1920s to provide large work-force pools for assembly lines that are no longer there.

As our city populations have grown larger and larger, as the city dwellers' apartment structures and office buildings haven gotten taller and taller, cities have become bustling metropolises that have expanded and ex-panded again into megalopolises. Over time, especially on the East Coast and West Coast along the oceans, these growing populations have eaten up greenways and greenbelts between cities, thereby becoming mega-mega-lopolises. This growth is happening throughout the world, while the internal land becomes more isolated and in many cases underused. Rural towns dry up and blow away like the deserted gold towns of the West. Some say that people occupy only 1% of the land. If people were spread out, everyone in the world would be able to have a comfortable home located within an area the size of the state of Texas. So why are we living on top of each other in highly congested areas? Why not formulate new cities using the new tech-nologies to help offset the pressures of everyday living in the mega-mega-lopolises? Why not actually attend a baseball game, football game, or tennis match in a home entertainment center, which functions as a residential full-screen, full-circle theater? Why not, via a high-definition travel log, actually experience the sounds, smells, and sights of visiting a foreign-city market-place? Why not browse through newspapers and magazines electronically and use cross-indexes to find further details, history, or reference works?

Do we really believe that radiologists will be happy reviewing only a few low-quality X-rays? Ask them! First they simply wanted a way to see a few X-rays to make a guarded determination. They need to quickly deter-mine if a Saturday-afternoon skier at a remote village needs to be taken to a big-city hospital for a broken leg, or is it simply a sprain? As time pro-gresses, they will next want to be able to get a second opinion by having another specialist in a distant hospital review a high-quality X-ray. Then they will want to be able to remotely control the type of X-rays taken and zero in on the area that they are most interested in, accomplishing all this from their home or office. Next, as they work with other technologies such as the MRI "donut," they might want to have more than 100 readings of a specific area. In the future, using sound waves, they will be able to see three-dimensional views of the head, which will require multiple readings and high-powered computer programs. With an ever-increasing appetite for more and more information, customers will need more and more band-width available for their more and more sophisticated computer programs.

No matter where we look, we see lists and lists of data beginning to be presented as images, graphs, video pictures, and sequential video displays in slow-motion, full-motion, high-resolution, high-definition, full-color form. This information will be processed, manipulated, and presented by computer-controlled display systems, large wall screens, robotic devices, sophisticated feedback/control systems, and real-time (live) operational systems. Information is obtained from sensors, polling devices, databases, historic files, radar systems, weather devices (noting sunlight, rain, snow, temperatures), pollution-measuring devices, earthquake-forecasting units, traffic measurements, pedestrian flow meters, heart and blood pressure monitors, environmental analyzers, etc. In the past, we have seen factories become more and more robotically automated with just-in-time inventory deliveries. Computer-aided manufacturing and computer-aided assembly have become fully achievable and linked with computer-aided engineering and computer-aided design packages. In the future, direct input from remote showrooms or customer phone calls and data calls will initiate the automated-service order processing and provisioning mechanisms as requirements are sent directly to worldwide automated assembly, packaging, and delivery systems. (See Fig. 4-1.)

Finally, as narrowband and wideband networks become successful and enable customers to store, access, browse, retrieve, concatenate, ma-

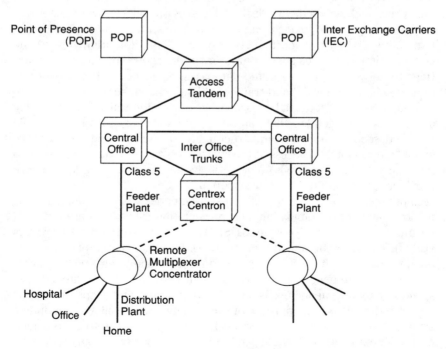

Fig. 4-1. Today.

nipulate, and present information, we will enter an era in which we electronically exchange letters, files, records, text, facsimiles, graphics, databases, books, and video conversations. Their success will require higher- and higher-speed switched broadband facilities to facilitate increased volume of lower speed narrowband and wideband information transfers. For example, local lanes and trails of a city will need to be interconnected by a superhighway as their traffic mounts. In time, the highways' access ramps and lanes will become contention points, requiring additional rings of the superhighway to be located closer to the vehicle's point of origin. So, as time passes, as traffic mounts, we must ensure that there are few congestion and blockage points that disable the narrowband and wideband networks. We must selectively shift their internetworking traffic to higher transport facilities. In time, over the next 50 years or so, access to unlimited traffic-handling facilities will be more closely available and accessible to the user. Similar to this local information growth, there will be a full range of narrowband/wideband/broadband offerings that will be integrated with the ever-expanding world of global communications.

Broadcast versus broadband

As we look at these differing new endeavors, we must be careful not to overextend offerings. For example, we must not think that a broadcast network, which is specifically designed to economically deliver one form of television-type video images, can easily become a fully interactive broadband network that transports high-resolution, high-definition videophone conversations, switches supercomputer traffic, allows numerous workstations to network with distributed databases, or facilitates the networking of narrowband and wideband traffic on higher-speed facilities.

Some have said it might be feasible, but it is impractical to initially deploy a broadcast arrangement that can be easily reconfigured later into a fully switched, secure, survivable infrastructure. A broadcast arrangement will not enable direct switched access at the shortest distance from the customer to interexchange carriers (IXCs). Nor will it give access to value-added network providers (VANs), alternative transport providers (ATPs), competitive access providers (CAPs), or specialized service nodes where additional information transport, manipulation, and presentation capabilities are provided by information service providers (ISPs) and enhanced service providers (ESPs). The future broadband competitive information marketplace must be open to everyone. It must be connected from anyone to anyone. To simply provide a one-to-many type of arrangement (broadcast) with limited low-speed any-to-any capabilities is ignoring the purpose and scope of a public broadband network. The purpose of such a network is to allow a continuing array of ever-increasing, value-added information services to be provided over the next century. A high-speed, ubiquitous, interactive, any-to-any offering is substantially different from simply extending current broadcast cable services. These "cable bell" video dial-tone-type services do not address the needs of the business community, government agencies, hospitals, or sophisticated work-at-home communities. These services offer just 500 one-way (downstream) broadcast channels and, on demand (via upstream data requests), access to selected movies, games, shopping catalogs, etc. This type of offering does not meet the need for a robust, secure, and survivable network that can handle voluminous, interactive, any-to-any traffic, nor does it facilitate private internetworking services.

There is a need to allow numerous service nodes to be overlaid and accessed by not only narrowband and wideband networks, but also by full broadband systems operating at 155 million bits per second. How will the deployment of these broadcast entertainment-based networks ensure that they serve as the network-of-networks infrastructure upon which global, national, and local networks actively internetwork voice, data, and video calls? Who is the keeper of the network? How do we ensure that the broadcast-based networks of substantially differing varieties are indeed compatible with the fully switched broadband ISDN networks being constructed on the standards of the International Telecommunication Union's (ITU) CCITT working groups? Will the streets be plowed up to deploy coax? Will areas where a few fibers are deployed for limited services be replowed to deploy more fiber when needed? Will facilities be dragged back to a central hub, or will new switching nodes be located closer to the customer? Will the network be constructed on fully switched, survivable rings based on SONET/SDH standards, providing access closer to the customer? Will these rings be able to handle 155-megabit customer traffic or only interactive, low-speed transport?

Yes, broadcast services providing entertainment will indeed be a subset of broadband services, but not the other way around. Interactive broadband services are not a subset of broadcast network topologies. When implemented within this broadcast arrangement, fully switched interactive broadband services are very limited and not easily growable. So with broadcast, we do not have a baseline network that can be easily and economically changed and expanded. What is required is to do it the old-fashioned way. When an industry such as transportation needed an entirely new airplane—such as the 747 or even the Concord—it started over from scratch. It considered all the previous designs, but it recognized the need for a substantially different, new structure that economically took advantage of the new technologies to obtain the new requirements. So it is with any new endeavor, especially when these new requirements are radically different from present endeavors. If indeed the industry is simply attempting to provide a parallel cable offering with more channels and a few upstream services, and that's it, so be it. This is not really a radically new venture requiring substantial redesign and change. We must realize that it can be easily superseded at a later time by some other network that takes advantage of the new technology. However, if this meets a specific, limited market segment, and this is the only market segment that the FCC wants the RBOCs to address via the "dial-up video" ruling, then this is acceptable. (But call it what it is. Don't call it something that it isn't.) However, if indeed the desire is to take advantage of the new technology and construct a new, ever-expanding facility of new services, it is time to establish a new, far-reaching base for going forward into the information era. This will be a public, fully switched, interactive, broadband information network that cuts across market sectors and has a new transport/switch distribution plant topology. It should be designed to meet the full range of needs of the information users—the telecommunications customers of the twenty-first century.

Broadband user needs

The future users of broadband information networks will require an exciting new array of features and services that meet their expanding information-handling needs and expectations. They will need to:

- Have fully interactive, high-quality videoconferences operating in the high-definition range and using large wall-mounted screen terminals that require information in the compressed form at rates of 50+ million bits per second.
- Have high-quality videoconferences operating with the highest resolution quality provided at rates of 50–155 million bits per second.
- Subsequently have wraparound, full-screen videoconferencing facilities requiring information transfer rates of 155–620 million bits per second.
- In time, have wraparound videoconferencing with holographic imaging capability operating at 620 million bits per second to 4.8 gigabits and higher.
- Enable supercomputer-to-supercomputer internetworking.
- Enable remote workstations to access fully distributed databases.
- Have fully interactive, fully switchable, fully dynamic bandwidth capabilities for "any-to-any" application.
- Ensure privacy, security, and survivability.
- Provide network access from locations situated close to the information customer, and from these locations to common carriers (IXCs) and alternative network providers (VANs, CAPs, and ATPs).
- Enable access to information service providers (ISPs), enhanced service providers (ESPs), database sources (DBSs), and global information sources (GISs).
- Provide internetworking capabilities for private-to-public-to-private information exchange from/to customer-premises networks, special-purpose networks, and shared private networks.

- Ensure that layered networks' layered services are interoperative and selectable by the customer.
- Ensure that narrowband and wideband networks remain versatile, parallel entities and become subsets of the overall broadband network to facilitate traffic flow and information interexchange.
- Ensure interoperability across narrowband/wideband/broadband services such as the various narrowband/wideband/broadband videophone offerings.
- Have addressability for every terminal or operating device, be it a PC or control system, where narrowband and wideband addresses become subsets of an overall broadband addressing scheme so consistency is maintained.
- Achieve simultaneous multimedia voice, data, video, image, text, and graphic information interexchange.

These needs will be constantly evolving and changing over the twenty-first century. They will require a fully supportable, fully expandable, fully interactive information infrastructure that is robust, secure, survivable, addressable, and ubiquitously available. It must interface to not only local, regional, and national networks and their service centers, but also to global entities via both wireline and wireless transports. Information will be transported at increasing speeds and increasing (volumes), growing from megabits (10^6) to terabits (10^{12}).

Market opportunity

These information customer (user) needs translate into considerable market opportunities. Key to these endeavors is a logical fiber deployment plan throughout the twenty-first century. It is obvious that entertainment has been already proven to be a segmented opportunity if deployed in such a manner as to encompass a wide area. Wireless technologies have already positioned themselves for providing economic area-wide penetration via microwave to high radio-tower transmission systems and via direct broadcast satellite arrangements. There is an opportunity for selected access to pay-as-you-go arrangements such as sporting events, games, movies, and other forms of entertainment. Besides electronically accessing these files, there is increasing pressure for using various competitive alternatives—such as movies delivered by pizza carriers—so no one will corner the market. It will be a wide-open game.

However, there is a singular, prime opportunity for electronic interactive services. Even leaders of direct broadcast satellite enterprises recognize and acknowledge the power of fully interactive services, different from their broadcast, selected-broadcast, or broadcast-on-request arrangements. The fully switched, fully addressable, fully interactive telephone, data-phone, and videophone offerings have a key role to play. Besides the videophone and videoconference opportunities, both locally and globally, there will be numbers of dataphone and imagephone broadband-type offerings. Computers will exchange large blocks of data among graphic/image systems that store and retrieve them from image databases for workstations that interactively design, change, and modify sequential video files to achieve high-quality video views of new architectural designs or video simulation models. These simulation models provide exciting views of changing environments, financial or economic conclusions, social conditions, geographical topologies, business opportunities, transportation flows, etc. Simulation programs will display complex relationships and interrelationships and move the user into the world of virtual reality. This will remove the barriers of conceptual thinking and provide an unlimited arena upon which to walk around the idea, problem, or situation and view it from any perspective.

The impact of interactive multimedia services has not been totally understood and appreciated. This newly budding tool with its simultaneous access and manipulation of data, images, graphs, video, and stereo-level audio information will require sophisticated information-handling mechanisms. As images become animated, graphs become three-dimensional, and large blocks of data are processed and reconfigured into visual entities, as more and more simulation programs establish virtual reality interplay, there will be an ever-increasing demand for greater and greater bandwidth. Similarly, as people repetitively perform functions such as reviewing visual files, they will express the need for faster and faster access and will be less and less tolerant of transport delay. So as we enter the world of public narrowband, wideband, and broadband information transport, customers will be continuing to push for greater and greater bandwidth as they attempt to accomplish more things faster and quicker. Soon they will forget earlier restraints of having to do things manually. How quickly those who drive automobiles with automatic transmissions forget the inconveniences of constantly shifting up and down as they turn corners or enter a freeway. Today, cars are not only considered for their transport capabilities, but also how many other creature-comfort features they contain to keep drivers happy. So as we enable the mind to more easily access information, it will soon want easier and easier and faster and faster access to more and more information. (See Fig. 4-2.)

As the narrowband/wideband offerings become more and more successful, their overall volume of transport will increase and increase again. This will require greater broadband facilities to internally handle this load without inhibiting throughput or creating blockage or congestion. Hence, we have moved the information user to an exciting new frontier—a fully interactive multimedia environment having voice, data, and video offerings. As this also encompasses broadcast capabilities, we have not only increased the number of television channels, but we also have increased the abilities of these channels to become high-resolution, high-definition of-

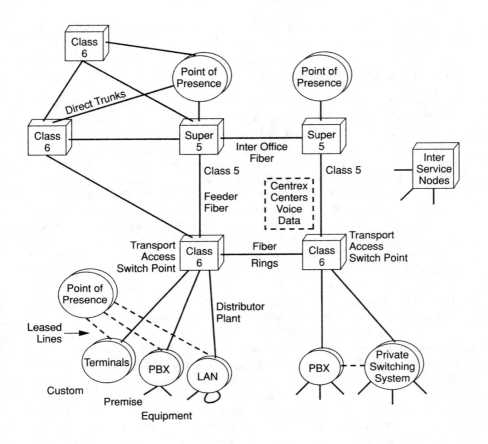

Public Basic Transport Company PBTC

Fig. 4-2. Tomorrow.

ferings rather than singular "me-too" cable offerings or simply the addition of more channels with some dial-up data services. We have changed the way we work, study, and play, but we have not limited our world to only transporting information in one-way, high-speed (downstream) vehicles with only low-speed (upstream) returns. We have achieved a fully interactive, high-speed interplay, one in which capabilities are not really limited. As the coax stops at the megabits, fiber continues to the never-ending terabits. This type of infrastructure—and the services its nodes provide— offer the marketplace an opportunity to meet the ever-expanding needs of the users throughout the next millennium, the information millennium. So what are the technologies that can make this happen?

Technical possibilities

After reviewing the applications, customer needs, and market opportunities, it should now be clear that the future public broadband network will be fiber-based and will require sophisticated switching systems that can adequately handle megabit to terabit traffic (see *Global Telecommunications*, Heldman, McGraw-Hill, 1992; *Future Telecommunications*, Heldman, McGraw-Hill, 1993; and *Information Telecommunications*, Heldman, McGraw-Hill, 1994). The broadband network will facilitate the movement of large blocks of interactive information among workstations, supercomputers, and databases. It will enable interactive, high-definition, high-resolution videophone and videoconference communications, and it will support high-definition broadcast information (selectively requested, periodically transported, or programmed for delivery). It will be able to integrate narrowband and wideband transport mechanisms with broadband traffic, thereby not only handling data message delivery and LAN-to-LAN networking, but also narrowband and wideband videophone communications.

Interfaces, therefore, need to be extended to existing systems, both wireless and wireline, as well as to future enhancements to these systems, particularly for their data-handling capabilities. Similarly, various future fiber enhancements need to be anticipated and taken into account at the time of deployment (for example, the considerations needed to achieve fiber to the home, fiber to the office, fiber to the pedestal, or fiber to the curb). The transition from hybrid coax/twisted-pair arrangements to pure fiber from beginning to end must be anticipated. We also need to ensure that the appropriate protocols are in place to enable access to service nodes above the network, and we need to ensure that they are under direct customer control. We must also resolve issues of security, privacy, survivability, error-free transport, and the need to adequately handle videophone or dataphone traffic to ensure against blockage, delayed delivery, and extreme congestion. All these and many more requirements at the bit and byte level of information transport present a complex packaging environment for the numerous types and forms of broadband traffic.

To accommodate these requirements, let us first begin by noting that we need to situate and locate the wideband network in an appropriate manner to later obtain the more sophisticated and diverse broadband offerings. We need to provide a new topology of survivable, switchable rings carrying traffic from one switching node to another around the ring or vertically up the network to more versatile, fully information-based central offices at the class 5 level of the network switching hierarchy. Digital data and video switching capabilities are overlaid on the new digital voice switches (such as the 5ESS IDN/ISDN switch) using the latest circuit- and packet-switching technologies (STM/ATM). In time, these voice switches with additional data and video switching capabilities will be replaced by more fully integrated electronic/photonic systems called superswitches. In a typical city having fifty or so voice-grade switching systems, there will most likely be only five to ten superswitches to serve the entire community. Eventually, these superswitches will use new wave-division-multiplexing (WDM) and space-division-multiplexing (SDM) technologies. The key to this architecture will be a new topology in which the new network access nodes become stand-alone front-end switches for these new superswitches. Thus, the new broadband switches come in two parts. One part is at these access nodes, located closer to the user and situated in small huts (protected, survivable, hardened) or collocated in small central offices. The other part, the superswitches, will be located in the larger metro offices or centrally located within the rural communities, providing integrated voice, data, and video switching and routing control.

Access to service nodes and application service centers can be obtained directly from the ring switches (network access switches) or from the more centrally located superswitches. Service nodes can provide a switching matrix to route calls to specialized service center units that provide such things as direct human operator data-call handling to achieve access to specific or specialized databases, language translation programs, specialized program compilers (syntax/semantics), complex global routing directories, address translators, etc. Alternatively, automated transport service nodes might include security coding capabilities (encryption) and other protective features for routing the video messages globally via satellite links, etc.

While the superswitches provide more network control, video switching, and data packaging, etc., the access nodes must interface to various protocols from/to private networks having internal asynchronous transfer mode (ATM) fast packet switches, or synchronous transfer mode (STM) variable channel circuit switches. Similarly, private-to-public wideband CPE to network-to-CPE interface arrangements are processed by these access nodes. In time, we need to add the additional capabilities for processing multiple wavelengths through the fiber. Scientists project that hundreds of differing bandwidths will shift fiber transport from gigabits to terabits. This new network topology will use survivable rings-on-rings architectures located closer and closer to customer premises with final subnodes extending the fiber facilities directly to the home or office. (See Fig. 4-3.)

In conclusion

This multilevel hierarchy with new switching nodes, which could be called class 6s, will in time interface to new CPE information PBXs (IPBX or IBXs), which could be called class 7s. These will have additional ATM/STM

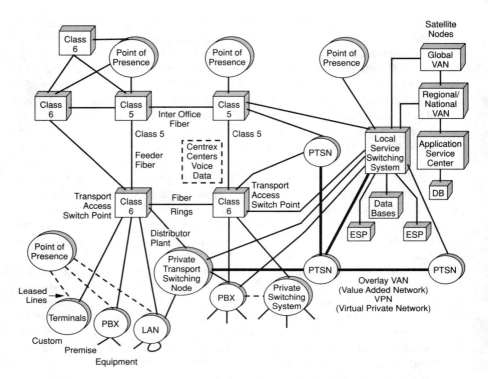

Fig. 4-3. The future.

data and video switching capabilities. These internal systems will be fully physically distributed through the customer complex, campus, or shared-tenant residential facilities. On the network side, service nodes can be reached from both the access switches and the superswitches. Similarly, both systems will offer direct paths to the IXCs, VANs, ATPs, etc. To this architecture, we can easily provide access to video databases via service nodes above the networks, using addressable switching arrangements. These access nodes will provide direct transport to the customers or to clusters of customers. Hybrid coax, twisted-pair, and fiber-optic arrangements will achieve the appropriate connections from the pedestal or curb to the home or office, and in time, these connections will become full fiber. Care must be taken not to engineer for a limited, blocked, concentrated, or contended neighborhood node due to limited fiber facilities. We must guard against designing the infrastructure just for the initial group of offerings. The whole spectrum of services must be considered at the time of physical plant deployment, thereby requiring the long-term view to achieve the correct supporting infrastructure. All in all, it is a new network, a new technical solution to a new market opportunity—the universally deployed public broadband information network.

5

Wireless information

"Vision is the art of seeing the invisible."
Jonathan Swift

As the Federal Communications Commission (FCC) releases more and more spectrum for wireless communication, which encourages a new family of services, perhaps some time in the future anyone will be able to talk to anyone from any location. Thus, it is time to look at the major issues and questions that need to be addressed and resolved, especially in light of the growth in the cellular field. (See Fig. 5-1.)

Cellular was originally a high-priced service for a limited number of mobile users who used expensive handsets and worked with limited spectrum. Personal communication services (PCS), based on a newly advocated low-power system, was to be the "Dick Tracy" watch for the masses—for the person on the street. It was to be priced as a low-cost service and provided via economically priced handsets. As government/military spectrum was being released for local commercial use, the intent was to have multiple providers offering these personal communication services within a local area. Wireless offerings were to be extended to form wireless PBXs within an office complex, enabling both voice and data to be easily transported around the office complex. As cellular handsets became smaller and less expensive, more people began using cellular. Cellular also went after some of the spectrum being made available for the PCS market, especially the 10-megahertz range, thereby extending their already-constructed infrastructure to reach more people. As cellular providers changed from using analog technologies to digital techniques such as time division multiplexing (TDMA) and extended TDMA, and on to spread-spectrum code division multiplexing (CDMA), they intended to not only better use their spectrum

2-WAY
WRIST
RADIO

Reg. U. S. Pat. Off.:
© 1962 by
The Chicago Tribune.

Fig. 5-1. Dick Tracy's two-way wristwatch. Tribune Media Services

and obtain more customers, but also to be able to make the conversations more secure. Hence, their handsets became smaller and easier to carry outside the automobile, and future changes will allow them to switch to and from cellular or PCS frequencies.

On the other hand, PCS requires an entirely new infrastructure in which its 30-foot antennas are located every 1000 feet or so and large, complex databases are required to keep track of a person's location. By combining paging techniques and enabling the return telepoint call via PCS, various forms of PCS can help facilitate services that do not require tracking the person. This has helped facilitate early operation in Europe.

As we review the differences between the macro-cell cellular and the micro-cell PCS entities and consider what the customer will pay for the services, we need to ask basic questions, such as: What percentage of the people will really pay to be found anywhere, anytime? Is the tracking system too complex? Does the paging system satisfy the requirement that customers can, when desired, selectively return phone calls via cellular phones, local pay phones, or personal outgoing service facilities? As we add advanced intelligent network (AIN) services to cellular, enhancing it and giving it more capabilities, we see that many of the differences between cellular and PCS are shrinking. However, PCS will work quite effectively inside an office complex or within a universally enclosed environment, where it's easier and less expensive to deploy equipment for a closed community-of-interest application.

As we look at data being transported at 8000 or 14,000 bits per second on a cellular system, specifically for mobile persons working out of their van, truck, or whatever, we need to consider cost. Hence, as cellular phone bills increase, many such users will rush to alternate means, especially when cellular services are priced quite high. So, in terms of both voice and data offerings, are there really some 60 million PCS users and 20 million cellular users who are willing to pay a high price for these services in order to have two types of networks that are quite independent of each other? Or are we really in an arena in which we selectively deploy and expand current cellular facilities and then overlay PCS capabilities in appropriate areas, using common handsets as we resolve the complex internetworking standards requirements for enabling multiple players. Indeed, it will be an exciting game as various wireless offerings are then tied to wireline facilities and internetworked. But first we must carefully consider: What do we want to provide? For whom? What will it cost? Will the perspective customers pay for it?

It is not the purpose of this analysis to pursue the wireless world in depth here. Such a detailed study is provided in appendix A. But it is important to suggest here that these networks are both complex and expensive. As efforts are spent to facilitate fiber to the home and fiber to the desk, it is indeed important to appropriately enable wireline-wireless access nodes interfaces. These will most likely occur at the class 6 and sub 6 locations serving the downtown areas, while cellular calls will most likely enter the network via traditional central office superswitch locations. Again, it is important to note that cellular capabilities will expand to encompass many of the PCS urban markets as cellular handsets get smaller and smaller and offer access to both frequency ranges. However, within office complexes and closed communities of interest (such as hospitals, university campuses, and entertainment centers), the very small PCS-type of wireless receiver will no doubt be viably achievable, perhaps in the size of the "Dick Tracy" watch. (See Figs. 5-2 and 5-3.)

Fig. 5-2. Calling Dick Tracy. Tribune Media Services

Fig. 5-3. Dick Tracy's TV watch. Tribune Media Services

See the excellent analysis on future wireless services provided by J. Hemmady in appendix A. He covers the expanding role of PCS, PCS's functional architecture, PCS's U.S. environment, cellular evolution to PCS, landline evolution to PCS, wireless data, wireless PBX, wireless Centrex, etc., as noted by the following abstract:

Personal communications services (PCS) are a set of capabilities that allow terminal mobility, personal mobility, and service mobility. The PCS concept is part of such initiatives as Universal Personal Telecommunication (UPT) promoted by standards bodies. PCS is generally segmented into lower-tier and high-tier services, significantly lower in cost than today's cellular service, providing significantly more mobility than today's landline service and having voice quality and feature sets similar to today's landline services. This analysis addresses how current communication networks might evolve in supporting the vision of personal communications networks (PCN), which is to create a single seamless network to provide services associated with a mobile individual or a personal identifier, and not simply a terminal. Specifically addressed are the evolution of two of the most common networks in the United States—the cellular and landline networks. Discussed are several new technologies that facilitate the creation of a PCN. Among the most important are SS7 signaling, intelligent network, digital switching, digital radio, portable terminal, and operations support technologies. Also addressed is an evolution of wireless communications to broadband networks.

6

The information highway to the information marketplace

*"Once there was an elephant
Who tried to use the telephant—
No! No! i mean an elephone
Who tried to use the telephone . . ."*
Laura Elizabeth Richards

The media's hype and hoopla over the early 1990 time frame for the information highway has been a little premature. The media have promoted such endeavors as: Internet's dial-up bulletin boards, which use analog, low-speed modems for voice-grade data traffic; Washington's nationwide MANs for interconnecting government research facilities and universities; or the FCC's push for telephone companies to expand cable's broadcast entertainment networks to include services on demand. These offerings are simply forerunners of what is yet to come. The real "information highway," which some call a "superhighway," is being defined by global standards committees such as the International Telecommunications Union's CCITT working groups. The seven-layer International Standard Organization's Open Systems Interconnection (OSI) model is being defined, overlaid, and augmented with numerous interface protocols for internetworking local, national, and global offerings that cover the full range of services, from physical transport interfaces to application presentation formats. The narrowband Integrated Services Digital Network (N-ISDN), wideband primary

ISDN (W-ISDN), and broadband ISDN (B-ISDN) transport structures are being finalized to ensure error-free exchange of global information. So let's review what we have learned concerning the forthcoming information highway to the information marketplace.

The information highway

What networks? What services? As previously discussed, there are three types of roads that formulate the fully interactive, integrated information highway. Using the existing copper-based plant, narrowband transport can achieve eight options of various voice and data traffic mixes, transporting information at rates of 64,000 or 128,000 bits per second with a parallel 16,000-bits-per-second path for signaling and low-speed packet-type data transport. We will see substantial growth in user acceptance in the local, state, and national arena once the network-terminating interface connections (such as the previously mentioned network terminating devices—NT2, NT1, and AT) become substantially reduced in price. This will occur as they are offered to larger and larger volumes of users (perhaps as $50–$100 interface chips) and when an economical narrowband ISDN network becomes ubiquitously available over the late 1990s and enables both circuit and packet transport (perhaps in the $25–$45 range).

This network will offer many exciting market opportunities. It will provide a factor-of-ten increase in relatively error-free, survivable data movement. It will offer security, privacy, and friendly access to distributed databases. It will also allow easy user access to information-service and enhanced-service providers' additional layers of services, such as: broadcast, delayed delivery, polling, sensing, record keeping, alarm monitoring, energy management, protocol interexchange, code conversion, access verification, authorization, and audit trails. Its global gateways will provide access to global application service centers and specialized global networks. In the same manner, local gateway servers and service nodes will enable access to private specialized networks for closed or specialized communities of interest, covering various industry applications such as health care and law enforcement.

Some have called this narrowband network the "starter" network or the "beginner" network that will initiate customer movement into the world of information exchange. This is indeed so because more usage begets more applications, which beget more usage. Indeed, there are many significant areas for immediate growth in narrowband data traffic: credit check, point of sale, database inquiry/response, data collection, and various data distribution applications for e-mail and facsimile messages.

However, as workstations become more sophisticated and cluster controllers group more and more terminal devices together, these vehicles will require a higher-speed highway. There will be a variable number of 64,000-

bits-per-second channels from the narrowband rates to 1.5 to 6.3 million bits per second, covering the full range of T1 and T2 copper plant, and on to fiber's 45-million-bits-per-second (T3) capabilities. These will provide the next level of fully switched, fully interactive services. In these switching arrangements, located on survivable rings positioned closer to the customer premises, we will see private-to-public-to-private access for local area networks (LANs) and customer-premises equipment (CPE). These will perform internal switching in the form of information private branch exchanges (called IBXs). The new asynchronous transfer mode (ATM) fast-packet technology and the synchronous transfer mode (STM) circuit-switching technology will come into play at the network node and within the customer office complex, university campus, and shared-tenant residential facilities. Indeed, using these variable, high-speed switching capabilities, instantaneous virtual networking will be achieved from one destination to another, thereby reducing the need for direct leased-line/trunk facilities. Key to these endeavors are the global data switching address capabilities, similar to those of the narrowband voice networks, to enable "any-to-any" switched data transport, both locally and globally.

Finally, greater transport volume will occur as the demand increases for more and more information interexchange in the form of visual images, graphs, and full-motion video. Then the need will emerge for SONET-based broadband transport. This form of superhighway will transport both narrowband and wideband traffic, as well as expanding new forms of visual information. Facilities must be deployed in such a manner as to ensure privacy and fully survivable, secure, error-free operation. Anything less would simply be an interim solution, for the more dependent the information marketplace becomes on its information highway, the more vulnerable its society. Hospitals and protection agencies will shift from point-to-point, leased-line facilities (which have 99.995%-reliable, error-free operation) to the fully shared, switched facilities. These shared, switchable arrangements, which facilitate easy, quick termination at different destinations and at variable rates, must also achieve (or better) the quality of the point-to-point networks. The switchable arrangements will require 99.999% or 100% error-free, survivable operation. (See Fig. 6-1.)

The information marketplace

Fully integrated multimedia services will cover the full spectrum of visual offerings, where images become animated, full-motion, and holographic in full-spectrum color, high-resolution, high-definition form. For this, we indeed will require the full "super" broadband capabilities of the information highway. Information will be in multiples of 50 million bits per second, reaching levels of 155 million to 620 million to 12 gigabits to terabits. This information will be transported over broadband facilities, as fiber passes

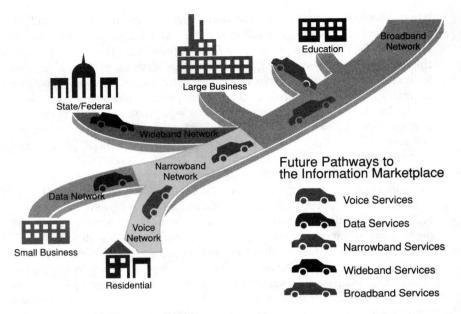

State/Federal
Large Business
Education
Broadband Network
Wideband Network
Narrowband Network
Data Network
Small Business
Voice Network
Residential

Future Pathways to the Information Marketplace

Voice Services
Data Services
Narrowband Services
Wideband Services
Broadband Services

Fig. 6-1. Future pathways.

more and more wavelengths using electronic/photonic switching and transport mechanisms. (See Fig. 6-2.)

Yes, indeed, we need to achieve a superhighway, but this roadway occurs in several forms: as a narrowband, wideband, and broadband passageway that enables fully switched, "any-to-any" interactive information movement. The information marketplace grows from the simplex point-of-sale, e-mail, fax, sensing, polling, and alarm-monitoring applications to the more sophisticated exchange of imaging and graphic displays, and on to the more complex, fully interactive multimedia arrangements. These will enable endless advancements and affect every aspect of society. In parallel, there is ample room for less demanding, separate wireline and wireless facilities that provide broadcast entertainment, audio telephone calls, and local area data networking. But as the next millennium develops, we will see that these early transport vehicles were simply the forerunners to the fully switched, fully interactive narrowband-wideband-broadband networks. These will provide the full range of information services to achieve the needed high-quality, high-resolution, high-definition, high-response capabilities for videophone, dataphone, imagephone, audiophone, and visionphone multimedia information exchange throughout the global information marketplace.

Narrowband Services

| Narrowband – Copper
64–128K b/s | Voice Messaging, CLASS, Selected Messaging Ckt Switching, Packet Switching, Protocol Conversion, Error Rated Control, Alternate Routing, Priority Messaging, E-Mail, Delayed Delivery, Voice/Text, Data Base Access, Encryption, Audit Trails, Broadcast, Polling . . . | Data Networks
Stock Exchange Network
Police/FBI Networks
State Agency Networks
Medical/Insurance Networks
Auto Parts Networks
Auto License Networks
Inventory Control Networks | Financial
Legal
Small Business
Residential |

Example Goals:
50% major cities by 1996
20% rural by 1997
70% urban by 1998

Wideband Services

| Wideband – Copper
– Fiber
128K-1.5M–45M b/s | Bandwidth Management, Wide Area Network, Dynamic Bandwidth, Private to Public Internetworking, ISDN-Non ISDN Internetworking, POP Access, Channel Switching, . . . Survivable Private/Public Information Internetworking | X-Rays, Patient Records
CAD/CAM Networks
Graphic Display Networks
Computer to Computer Networks
1.5M b/s Picturephone Networks
Video Conference Networks
Wide Area Networks | Medical
State
Manufacturing
Securities |

Example Goals:*
20% major cities by 1996
10% rural by 1997
50% urban by 1999

Broadband Services

| Broadband – Fiber
N (50M b/s) | Multiples of 50M b/s Switched Transport Video Conferencing, Picturephone, HDTV, Computer to Computer, High Speed Data Transfer, SMDS, FDDI, Frame Relay, Call Relay (ATM), & Broadband Information Transfer, Storage, Access, & Presentation | High-Resolution Picturephone
Entertainment
Media Events
Education Video
Medical Imaging
High Definition CAD/CAM
High Definition TV
Computer to Computer
Data Base Manipulation
Visual Presentations | Education
Entertainment
Large Business
Residential |

Example Goals:*
10% major cities by 1997
5% rural by 1998
20% urban by 1999

*These example goals are presented as an example of the need for having a plan of action that can be realistically achieved & agreed to by all parties. Each LEC must establish this type of program in order for suppliers & users to adequately prepare for these types of offerings.

Fig. 6-2. Goals and objectives.

Narrowband

Besides establishing a public digital voice network, it's interesting to see how several different types of narrowband data transport roads can be developed to meet the needs of differing transport vehicles. We have noted eight distinct options with specific basic network feature packages. These offerings can be enhanced and expanded by advanced intelligent network platforms and applications-service-center nodes. We have seen a whole host of new, exciting technical possibilities and market opportunities, but we have also noted that they must be priced on an overall basis (locally, regionally, nationally, and internationally) to ensure growth. We want to enable firms to exchange information locally and globally in a cost-effective manner, thereby eliminating various unneeded, energy-consuming modes of operation, such as physical travel. This achieves increased productivity and personal effectiveness, not only in the business place, but also in the home. Personal issues such as quality of life, education, health, social interaction, and entertainment are enhanced as highway congestion reduces, queues are eliminated, and rural homesteads blossom. Establishing a public narrowband data network is the first step. Some might say this is a baby step, but it is an essential step as we readily take advantage of existing plant to address 70–80% of today's information exchange application needs. With this step, the baby bells (and other local exchange carriers) can successfully enable their customers to enter the global information marketplace.

User needs

- Audio communications.
  - —Call control features.
  - —Specialized call control (based on calling party).
  - —Customized call control (CLASS).
  - —Extended call control (AIN).
  - —Stereo-quality audio.
- Data communications.
  - —Standard interfaces.
  - —Global addressing.
  - —Error-free transport.
  - —Fast call handling.
  - —Shared transport.
  - —Usage pricing.
  - —Security and privacy.
  - —Virtual private networking.
  - —Economical fast transport.
  - —CPE signaling, sensing, polling.
  - —Multiparty messaging.
- Video communications.
  - —Small screen (eye to eye).
  - —Economical quality imaging.

Network options
One

- 64-Kbps digitized voice network interface.
- 64-Kbps circuit-switched data network interface.
- 16-Kbps signaling/packet channel.

Two

- 64-Kbps circuit-switched data network interface.
- 64-Kbps circuit-switched data network interface.
- 16-Kbps signaling/packet channel.

Three

- 128-Kbps circuit-switched data network interface.
- 16-Kbps signaling/packet channel.

Four

- 64-Kbps digitized voice network interface.
- 64-Kbps packet-switched data network interface.
- 16-Kbps signaling/packet channel.

Five

- 64-Kbps packet-switched data network interface.
- 64-Kbps packet-switched data network interface.
- 16-Kbps signaling/packet channel.

Six

- 128-Kbps packet-switched data network interface.
- 16-Kbps signaling/packet channel.

Seven

- 64-Kbps circuit-switched data network interface.
- 64-Kbps packet-switched data network interface.
- 16-Kbps signaling/packet channel.

Eight

- Two 64-Kbps digitized voice network interfaces.
- 16-Kbps signaling/packet channel. (See Fig. 6-3.)

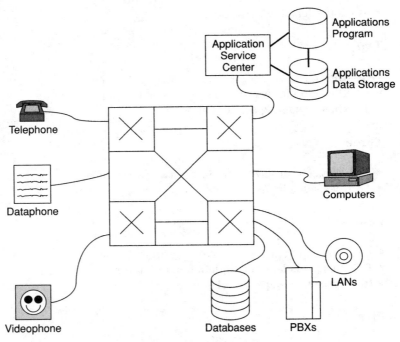

Fig. 6-3. Narrowband.

Network services

- Second audio line.
- Class services.
- Caller ID services.
- High-quality 7-kHz stereo.
- Integrated audio and data ISDN standard interfaces.
- Digitized CPE-to-CPE transport-security-privacy.
- Packetized data.
- Universal numbering/addressing.
- Closed user groups.
- Broadcasting.
- Priority override.
- Delayed delivery.
- Data caller identification.

- Polling/sensing.
- Error detection/correction.
- Encryption.
- Dynamic routing.
- Connection-oriented/connectionless transport.
- Data networking, protocol conversion, code conversion.
- Image transfer.
- Robust transport-survivable alternate routing.
- D-channel signaling.
- Data packet networking.
- Application-service-center accessibility.

Service applications
Voice

- Multiple lines.
- Small business, residence, etc.
- Audio stereo.

Data

- Point of sale.
- Alarm monitoring.
- Environmental control.
- Inventory control.
- Credit checking.
- Weather sensing.
- Home incarceration.
- Traffic monitoring.
- Automatic-teller control.
- Text messaging.
- Group 4 fax.
- X-ray transfer.
- E-mail.
- Global data directory.
- Data storage.
- Delayed delivery.
- Private-to-public networking.

Video

- Videophone (128 Kbps).
- Desktop videoconference.

Wideband

The wideband information network will be initially deployed as an overlay information offering for interactive voice, data, image, graphic, and video traffic. It will operate as fully switched multiples of 64,000-bits-per-second channels, up to T3 rates of 45 million bits per second. These transport offerings will eventually become virtual tributory subrates of the SONET OC-1 (50-million-bits-per-second) broadband information network. This wideband network will initially facilitate existing LAN-to-LAN internetwork traffic, but it will primarily provide fully switched, interactive facilities between workstations, mainframes, and databases. It will also enable supercomputer-to-supercomputer interconnect traffic and interactive wideband videophone and videoconference communications. This offering will be a prelude to the forthcoming fully interactive, fully deployed broadband information network.

User needs

- Variable bandwidth on demand.
- Variable destination addressing.
- Switched virtual private networking.
- Usage pricing.
- Error-free transport.
- Nonblockage.
- Noncongestion.
- Fast setup and take down.
- Dynamic reconfiguration.
- Full range of variable, high-speed wideband transport.
- Security/privacy.
- Survivability.
- Global accessibility.
- Universal availability.

Network options

- N × 64,000 bits per second.
- Range: 128 Kbps to 45 Mbps.
- Availability: standard, up to 1.54 Mbps (copper); customized up to 6.02 Mbps (copper); selective up to 45 Mbps (fiber).
- Primary-rate ISDN (switched) 23 B (64 Kbps) + D (64 Kbps). (Where each B channel service is identified via the D channel for integrated voice and data transport.) (See Fig. 6-4.)

Network services

- Global addressing.
- Virtual networking.

- Bandwidth on demand.
- Precall routing and bandwidth selection.
- Usage sharing.
- Variable/dynamic bandwidth.
- Error-free transport
- Security/privacy mechanisms.
- Survivability/alternate routing.
- Delayed delivery.
- Broadcasting.
- Sensing/polling.
- T1/T3 networking.
- Fractional T1/T3 networking.
- Primary-rate ISDN networking.
- Frame relay/SMDS networking.
- FDDI interfacing.
- Global service center access.

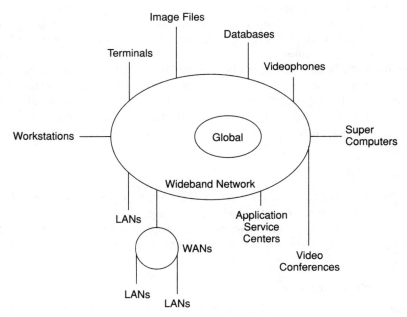

Fig. 6-4. Wideband.

Service applications

- Public WAN/MAN networking.
- LAN-to-LAN transport.
- Virtual private networking.

- Supercomputer-to-supercomputer transport.
- Workstation-to-supercomputer transport.
- Videophone, 384 Kbps/1.5/45 Mbps.
- Multimedia workstations networking.
- Private-to-public-to-private networking.
- Access to local, regional, national, and global databases.
- Customized application service centers.
- CPE-LEC access to IXCs, CAPs, ATPs, and VANs.

Broadband

The public broadband information network will address a never-ending, expanding family of customer needs for fully interactive voice, data, and video services. Supercomputers will internetwork among themselves and to distributed databases and will be accessed by sophisticated workstations to enable complex analyses that present solutions in the form of video images and graphic displays. Videophone conversations and video-conferences will require fully switched, interactive high-definition, high-resolution capabilities. Similarly, multimedia systems will require more and more information handling as these versatile tools become used in numerous applications throughout the marketplace. Upon these endeavors, broadcast capabilities will be overlaid to provide entertainment services. So, all in all, there is a need for an ever-expanding interactive, robust, survivable, secure, public broadband information network that facilitates the transition of today's industrial-revolution-based twentieth-century society into the information-based society of the twenty-first century.

User needs

Users need a full range of voice, data, text, image, video, and vision information monitoring, sensing, polling, auditing, exchanging, storing, accessing, listing, searching, browsing, retrieving, processing, manipulating, and presenting to any, anywhere, anytime, in any multimedia combination.

Network options

Network options include a full range of switched SONET transport as multiples of optical carrier-1 (OC-1) rate (50+ Mbps) with international interfacing at OC-3 (155 Mbps) and OC-12 (620 Mbps) from the customer premises, overlaid with HDTV broadcast capabilities. (See Fig. 6-5.)

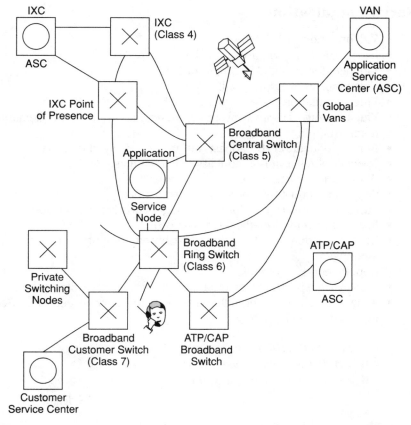

Fig. 6-5. Broadband.

Network services

- Broadband addressing.
- Global networking.
- N-W-B networking.
- Interactive switched broadband videophone (50/155 Mbps).
- N-W-B videophone integration.
- Multichannel broadcast television.
- Full range of wideband services extended to broadband rates.

Service applications

- Videophone (50/155 Mbps).
- High-definition TV/conventional TV.
- Videoconferencing (155 Mbps) or less.
- Video picture windows, wall-size travel logs.
- Video simulation systems, virtual reality.
- Video storage retrieval systems, video files, movies on demand, home shopping/catalogs.
- Remote education (broadcast, interactive, delayed delivery).
- Specialized/customized educational video files.
- Workstation access to remote databases.
- Supercomputer-to-supercomputer transport.
- Supercomputer-to-distributed-intelligent-workstations transport.
- Imaging systems—medical tests/records.
- Global broadband networking, distributed computing, holographic meetings.
- Robotic control systems.
- Energy/transport/defense online feedback-control systems.
- Access to universal broadband application service centers. (See Figs. 6-6 and 6-7.)

"Where is the life we have lost in living?
Where is the wisdom we have lost in knowledge?
Where is the knowledge we have lost in information?"

T.S. Elliot

#1 Voice Network	
#1 Voice Network +CCITT #7	Public Voice Networks
#2 Special Circuits LANS/MANS	Private Transport Networks
#3 ISDN Basic "B or 2B" Circuit Data	
#4 ISDN Basic "B or 2B" Packet Data	Public Data Networks Narrowband
#5 ISDN Basic "D" Channel Services	
#6 ISDN Primary (23 B) (Switched)	
#7 ISDN Primary (23 B) (Non Switched)	Public Wideband Networks
#8 Broadband Channel Switched Transport Access Nodal Rings	
#9 Broadband Switched	Public/Private Broadband Networks
#10 CPE Broadband Switched	
#11 Interexchange Carrier	Common Carrier Networks
#12 Global VAN	Private Value Added Networks

Fig. 6-6. Information networks.

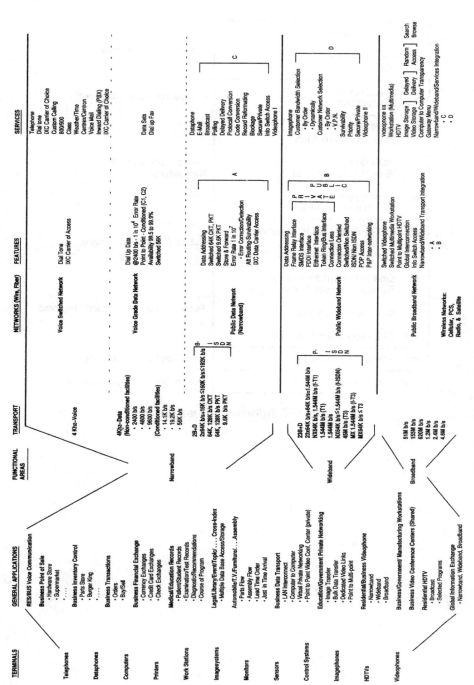

Fig. 6-7. Application terminals to services.

Part 2

Plan for the future

*"Ours is too busy a world,
and there is not time . . .
. . . for considering."*

Louis L'Amour

7
Issues and strategies

We do not plan to fail . . .
We fail to plan.

As we consider the various aspects of our vision for the future and appreciate its subtleties, we need to construct an efficient and effective plan that covers the key issues and strategies and addresses the following questions: How can we achieve the vision? When? What path do we take? What steps do we make? What tasks do we perform in anticipation, in preparation, in sequence, in parallel? (See Fig. 7-1.)

Analysis

As we pursue our plan for the future, let's consider the following issues: complexity, uncertainty, cost, long-term revenues, and risk. Today there are few easy decisions. We now have to twist and turn many issues as we pursue the various technical possibilities and market opportunities. We must not only address the financial aspects of cost and short-term versus long-term revenues, but also the public policy and people issues, and the negative and positive effects of many of the new technologies on society.

As we pursue individual pieces of this complex puzzle, it becomes more and more evident that they are indeed all interrelated. Just like attempting to solve the Rubix Cube, as we twist and turn a particular market service, we need to position its technology and financial aspects into a more complete offering, for it must be integrated with the other technical offerings, financial considerations, and management expectations. Many times these aspects are diverging rather than coming together into a consistent, coalesced offering.

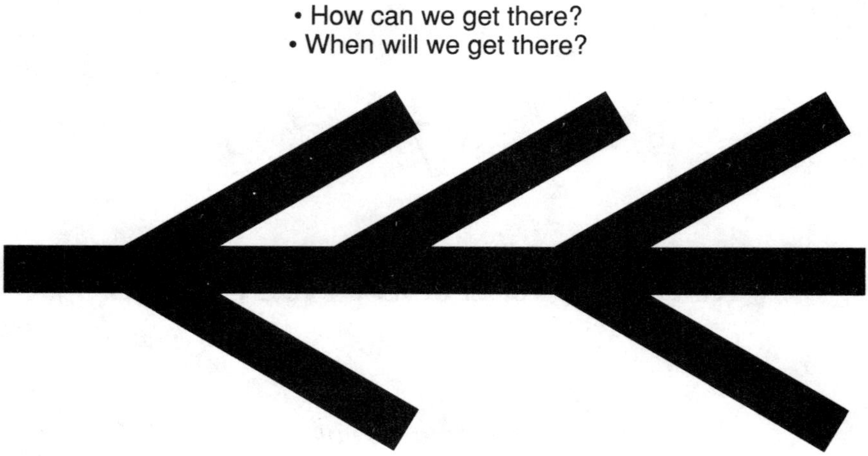

• How can we get there?
• When will we get there?

Fig. 7-1. Plan for the future.

Neither the computer industry nor the communications industry can continue to downsize without new offerings. As we continue to shrink around existing services, we all begin competing for the same target market, but with dwindling share as more and more global players enter the game. As our focus narrows, as we concentrate our endeavors into singular, more effective, more specific offerings, many times we leave widening holes for the new players to enter. As new technology pushes and the market pulls, these opportunities wait for no one as they continue to expand an ever-increasing information marketplace. Hence, if we simply perform a better job on voice and miss the ever-expanding data and video market, we might indeed find ourselves far from being able to provide integrated voice, data, text, and video multimedia services.

Today, as never before, we have an abundance of new technology that allows us to participate in multiple, fully switched, interactive arenas, specifically narrowband, wideband, and broadband information services, as well as broadcast entertainment and wireless.

It is important to address all these opportunities, realizing that expenditures can be huge and many revenues are long term in coming. We need to develop and nurture new markets for balance sheets to shift from red to black. We need long-term staying power. For example, *USA Today's* endeavor took 10 years and $1 billion to gain and establish a mature market position to move it from the red to the black. Fortunately, or unfortunately, depending upon one's perspective, the "baby bells" have had a similar period of opportunity with billions to spend. However, ten years after divestiture, few RBOCs have new offerings that have successfully increased revenues, other than simply moving existing customers from one technology to another (such as leased-line data transport to shared-usage technologies).

In this regard, market tests have shown the customers' desire for a fully interactive, publicly switched data network that is universally available and offers easy, standard communications that allow terminals of differing types to easily address computers of differing types, as well as each other. It appears that telephone companies need to offer an increasing amount of transport capacity at economical rates to encourage usage so hosts of new, varying transport-dependent services can flourish and blossom. So what should be done to create sufficient demand to encourage quick initial acceptance by the customer? What is needed is massive advertising to explain that a new public data network is available and show the customers how to use it. That way, their initial usage encourages more usage, fostering their need for more transport capacity. How should data services be priced to encourage this growth? How ubiquitous should offerings be from ocean to ocean? How are all these ventures financed—simultaneously or phased? How are our quicker and faster standards for CPE-to-network interfaces, switch to switch, and network to network, achieved to encourage private and public internetworking? Is the world a continuum of LANs or a shift to fully, publicly switched facilities?

To accomplish early market penetration and growth, providers need to change the work force's mode of operation. Providers need to educate their customers on how they can use cheap, economical, instantaneous communications to modify their personal air travel, overnight document delivery, automobile trips to local meetings, etc. To do so, they need to enable inexpensive videophone, videoconference, and secure electronic data transport services to be available throughout selected areas and not in isolated pockets. The greater the area, the greater the usage. For example, it was found in Japan that once ISDN was ubiquitously available, its usage grew exponentially.

Hence, we can better use our natural resources and remove some of the congestion in megalopolises, allowing people to work in satellite areas on the rings around the major cities, or at home, or in a rural environment. This would achieve a happier work force that is better able to cope with the pressures and complexities of business, as well as provide a solid and comfortable family environment, thereby increasing U.S. work force productivity and global competitiveness.

Pricing

Providers and suppliers can no longer price new offerings high and wait for slow market acceptance to increase to a point of bringing prices down. They need to aggressively deploy new products priced for encouraging mass market acceptance. It does little good to have a multibillion dollar network established for a few video users who are waiting to talk to a few more video users. They need to price services independently of transport bandwidth, on a service basis, especially as transport capabilities increase. (See appendix B.)

Providers need to use existing copper plant to better transport data and video and then selectively deploy fiber. Cost numbers to refiber America are in the $200–300 billion range. There are many applications (some say 80%) that can be economically served by a new, fully switched, fully interactive, copper-based public data network at narrowband ISDN rates of 64,000 and 128,000 bits per second.

Providers can then selectively condition copper to enable wideband rates of $n \times 64,000$ bits per second to dynamically move data up to T2 rates of 6.3 million bits per second, and then selectively deploy fiber to handle T3 rates of 45 million bits per second.

Finally, they can go to the full fiber-based broadband, using SONET/SDH (Synchronous Optical Network/Synchronous Digital Hierarchy, American/European Digital Hierarchies) optical carrier rates. Multiples of 50 million bits per second are switched at internationally agreed customer interfaces of 155 and 620 million bits per second. These offerings are served by 2.4, 4.8, etc. gigabit (10^9) transport rings, which will subsequently operate at the turn of the century at terabit rates (10^{12}) to meet the needs of high-definition television and videoconferencing, as well as distributed supercomputer data communication and imaging systems. (See Fig. 7-2.)

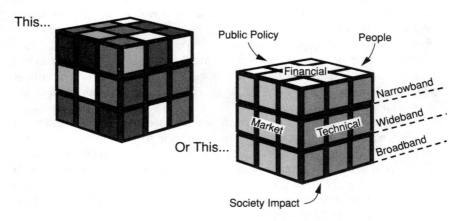

Fig. 7-2. Managing the information game.

So, how do we differentiate these narrowband, wideband, and broadband technical possibilities, with their hundreds of future services for thousands of applications? Which one or ones should we provide for today's and tomorrow's marketplace?

Where should we put our full resources? What is best for society? What is best to make the U.S. work force competitive in a global economy?

How can telecommunications enable the computer industry to enter every facet of the marketplace?

What should be done today, now that ISDN's narrowband standards are more fully resolved and in place in chip sets? Do we build a switched, interactive public network that can be easily accessed and economically used? With such a network, patients' records can be quickly moved, remote information databases can be accessed, images can appear on multimedia terminals, narrowband videophone conferences can be held in the 4- to 5-inch matrix form, and every terminal can be easily accessed and addressed and interconnected at 128,000 bits per second. That's 10 to 20 times faster than analog voice-grade transport rates, where error rates shift from 1 error in 10^4 to 1 error in 10^7 or 10^{11}. The public data network would provide secure and efficient sensing, broadcast, polling, routing, and survivable transport to enable computers to move from the accounting/financial services arena to online point of sale, real-estate listings, medical X-rays, inventory control, product distribution, and international business activities—or should we wait for broadband fiber to be deployed throughout the United States?

How about wideband? Should telephone companies simply modify their copper's capacity to meet the needs of expanding workstation requirements and LAN interconnects, or should we concentrate on broadband/broadcast endeavors such as dial-a-movie, home-shopping, and play-a-game? Is it appropriate to plow up a street to deploy coax, or should they go to full fiber deployment? Do the providers have the resources to do it twice—once in the 1990s and then again in the early twenty-first century? Should they build a switched broadband network so high-quality interactive videophone and videoconference services prevail? Should these new networks be available to enable distributed computers of increasing power to have switched, high-speed access to large mainframes, controlling expanding databases, and having supercomputer offerings? This would enable data to move from the alphanumeric world of text, to still frame images, to fully dynamic multicolor, high-definition video. Where and when should money be spent? The media's hype of the RBOC-cable company ventures are increasing demands for immediate entertainment offerings, but what about the computer industry? It is under serious financial stress and needs to expand its offerings universally, but it requires data networks that enable "many-to-many" or "any-to-any," instead of only "one-to-many" services.

So as we review what's happening, where we are going, and where we go from here, we need to determine how we can best work together to achieve new supplier product lines (requiring billion dollar expenditures) that will be used by network providers (requiring multibillion dollar expenditures) to enable families of new CPE terminals to provide information services to families of customers across the United States.

As we go from ideas to products, we need a totally new model that enables us to better achieve our destinies. Some do not want to initially change what they do but how they do it, but we all know that this evolutionary change in our mode of operation leads to a revolutionary change in what we do.

During the new millennium, information communications customers will need totally new services to meet their personal, business, and family needs. Providers will need to meet this challenge by using the existing and new fiber-based plant to provide a growing array of new evolutionary and revolutionary narrowband-wideband-broadband information services.

So what are the information communications users' needs? What is the market? The applications? What is the opportunity? What are the issues? What paths do we pursue? What steps should we take? But first, what are the needs that must be fulfilled?

Information customer needs

- • Control information movement and services.
- • Rapidly exchange information.
- • Access remote databases.
- • Translate data into graphics.
- • Store and retrieve images.
- • Dynamically move varying amounts of information, requiring dynamically changeable transport capabilities.
- • Increase personal knowledge and skills.
- • Reduce stress and increase performance.
- • Manage time.
- • Access information on a timely basis.
- • Tie dispersed communities of interest together.
- • Enable small firms to compete competitively.
- • Enable large firms to improve performance and productivity.
- • Reduce travel time.
- • Access and exchange information.
- • Improve quality and timeliness of decision making.
- • Enable the internetworking of private and public networks to encourage usage and growth of shared public facilities.
- • Enable interoperability of computers of different capabilities.
- • Provide security and survivability of information transport throughout the network.
- • See information:
  —In text form.
  —In image form.
  —In graphic form.
  —In video form.
  —In person.

- Enable our customers to be successful in:
  —Business.
  —Family life.
  —Personal life.
- Improve the quality of life.

Market needs

The market needs a structure that enables:

- New services—both the regulated and nonregulated, in a timely fashion.
- Enhanced voice services.
- Data, image, and video services.
- Interoperability of computers of different capabilities.
- Internetworking of private and public networks to encourage usage and growth of shared public facilities.
- Secure and survivable (S&S) transport of information.
- Growth of new services as features are added to features to meet the new and changing needs of our customers.

Network needs

- Use our existing copper facilities to the maximum to provide new data services over the narrowband 64-Kbps range of offerings.
- Enhance our existing copper facilities to the maximum to provide new data services in the wideband ($n \times 64$ Kbps from 128K to 6.3 Mbps to 45 Mbps (fiber)) range of offerings.
- Establish a new fiber-based structure from which to offer exciting and revolutionary new graphic, image, and video services in the broadband ($n \times 50$ Mbps) range of offerings.
- Provide the necessary support systems to ensure operational success from the customers' provisionary, deployment, and delivery perspectives.
- Ensure ease of growth of information movement:
  —More and more usage.
  —Higher and higher speed.
  —No bottlenecks.
  —Narrowband, wideband, broadband.

Management needs

- Formulate both an evolutionary and revolutionary infrastructure for supporting and expanding voice/audio services, as well as extending to encompass narrowband, wideband, and broadband data, image, and video services by the turn of the century.

- Be able to evolve new features in a timely manner from previous features as user needs become more and more complex and extensive.
- Achieve this infrastructure for specific markets, as well as for general markets for more ubiquitous applications.
- Obtain an automated self-diagnosing, self-recovering, self-healing infrastructure that enables less-sophisticated technicians to successfully manage more complex technology.

These transport, market, and network needs require a layered networks' layered services infrastructure that supplies narrowband services over existing facilities. These would be overlaid with conditioned wideband services and then augmented by a growing array of broadband services to formulate an expanding and encompassing fiber-based broadband network services infrastructure. Hence, there is a need to establish, in the 1990s, the correct infrastructure to enable U.S. firms to successfully participate in this exciting information arena.

To help focus this vision and obtain this supporting infrastructure with its revenue-producing new services, twelve strategic challenges (goals) are identified that need to be achieved by the end of this century, the end of this millennium, in order to be successful in the next. (See Figs. 7-3 and 7-4.)

Our first challenge

Providers and suppliers need to establish ISDN access interfaces to numerous information services. ISDN has been called the Integrated Services Digital Network. It provides seven levels of interconnect standards over the customer-network interface to enable the movement of voice, data, image, and video information to and from users' terminals, telephones, televisions, computers, environmental control systems, or workstations.

Initially, ISDN interfaces will be used as vehicles to access the new digital public data networks, both low and medium speed, over narrowband ISDN interfaces of 64,000 bits per second (basic rate), up to its wideband 1.5 million bits per second (primary rate). Later in the 1990s, multiples of 50 megabits per second of data—up to ten or so gigabits—will be the norm. These superfast, high-speed transport interfaces will be provided by broadband ISDN.

Therefore, we need to aggressively provide these initial low- to medium-speed ISDN access interfaces to every data user within the United States over the 1990s. This access must be priced to encourage usage and growth. Seven out of ten data users today will be satisfied by narrowband basic rate interface BRI-ISDN, and two out of the remaining three will be satisfied with wideband primary rate interface PRI-ISDN. But as usage and maturity increases, all will desire more and more data manipulation and presentation capability, as well as transport transparency, especially as

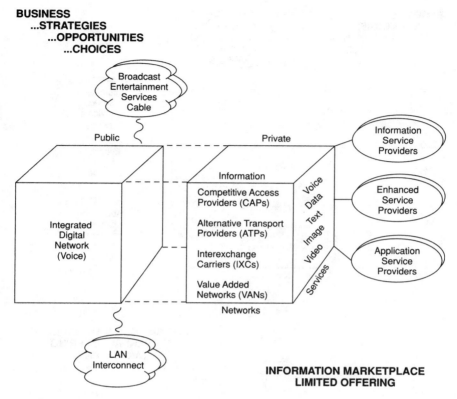

BUSINESS
...STRATEGIES
...OPPORTUNITIES
...CHOICES

Fig. 7-3. This.

users demand 50-megabit videophone services. To meet these expanding needs, broadband ISDN interfaces should be available by the mid to late 1990s. We have seen the computer power/computer usage/greater computer power/greater computer usage cycle continue until current computer capabilities and usage are doubling every 18 months. Similarly, private communication network capabilities have grown from 1 million to 100 million bits per second. So it will be with new distributed information services. Their growth must be fostered by providing private networks with wideband/broadband conduit to publicly facilitate these new information services via the new frame relay/SMDS/FDDI/B-ISDN access interfaces.

Challenge two

Public network providers need to continue their U.S. infrastructure digital upgrade program by using current host/remote (or sometimes called base/satellite) distributed switching technology in the rural/suburban en-

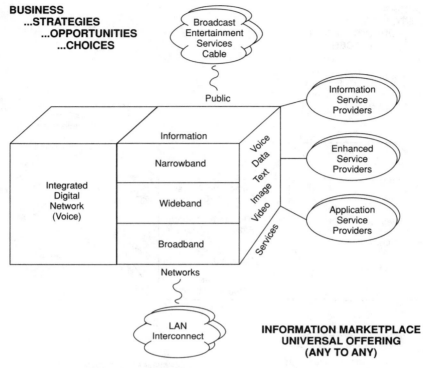

Fig. 7-4. Or this.

vironment. A digital host (base) unit can be placed at the county seat to simultaneously upgrade 15 to 20 small towns in the vicinity using remote satellite units. Using these digital capabilities, maintenance and support operations can be automated. Where appropriate, digital bases can be co-located next to analog systems in urban environments to enable business centers to access digital ISDN Centrex-type services for new offerings such as voice mail.

Similarly, they can access digital data networks by using the N-ISDN data interfaces provided from these remote switch units. As part of this upgrade, new signaling systems, such as International CCITT SS7 Common Channel Signaling and CCITT ISDN "D" Channel Signaling, will not only speed the calls through the network, but will also enable numerous new services based upon calling party identification. These services could include selected call transfer, call screening, selected messages, selected intercept, delayed delivery or reroute, and new phones and terminals that inform the customer exactly who is calling by name or number. Alarm monitoring, energy management, and environmental control systems will also use the new "D" channel transport services. In years to come, terminals will use these signaling systems to obtain new features and services from the network for yet-unknown applications. Private networks will need standard interfaces to the public network infrastructure to obtain private/public (P&P) internetworking.

Challenge three

Providers need to overlay a public data network using narrowband/wideband ISDN interfaces to provide ubiquitous, low-speed (less than 64 Kbps) to medium-speed (64K/128 Kbps) to high-speed (up to 45 Mbps) transport in a shared-data-packet or dedicated-circuit-switched mode of operation. We will see the transition from analog to digital for user interfaces to the data network. To stimulate early removal of analog devices from the dial-up voice network and all the delivery problems they incur, change is needed from flat rates to usage-message rates. To encourage this, RBOCs and VANs will need to provide the full spectrum of data transport services, from error detection and correction with alternative routing to message broadcast and data polling. This will enable the successful entrance of new services such as datafax, electronic mail, digital music transfer, remote medical treatment, home incarceration, weather monitoring, financial services, inquiry/response centers, and work at home.

Challenge four

Providers and suppliers need to establish a public and private network fiber deployment topology plan to extend fiber transport capabilities to home and business customers. Once this plan is in place, we can then begin implementation in the mid-1990s to "rewire America" with fiber. The completion target for restructuring all major wire centers serving both small and large business communities, as well as selected residential areas, will most likely be the second decade of the new century (2015 is the proposed congressional target for broadband network deployment). As part of this strategy, transport-access switching centers should be positioned closer to the user to provide alternate routes for survivability and open-network-architecture/open-systems interface points for enabling private-to-public internetworking.

This allows for new services for bridging and routing local area networks together over the public network, as well as for providing LAN gateway access to private MANs, WANs, and POPs (points of presence) of national network providers. These include MCI, US Sprint, and AT&T, who interconnect business campuses. In this manner, competitive access nodes are established closer to the customer to set the stage for future regulatory concurrence that will enable full "information networking" of "information services."

Challenge five

Private and public network providers need to overlay broadband switching service capabilities on the digital network. The first step is to perform trials using suppliers' technologies to establish the new service requirements needed in the next-generation systems that are destined for the late 1990s urban environment. The suppliers and providers must determine (from both a technical and market perspective) the correct level of integration, synchronization, and mixture of bursty variable-bit-rate (VBR) or long continuous-bit-rate (CBR) voice, data, and video information that needs to be transported for various types of customers. Specifically, numerous ways must be explored for providing video images for the different applications in order to identify the types of video services that will become the most successful in the new era, and channel broadband services in the desired direction.

The trials must sift through the ever-changing arrays of opportunities, from still-frozen images, to slow frame, to talking heads, as well as to commercial TV, advanced TV, and high-definition TV. We jump from 9600 to 64K to 384K, and on to 1.5M, 6.3M, 45M, 135–155M, and 620 megabits while using various compression and enhanced imaging type transmission techniques. Similarly, we must understand and identify megabit switched data services, that enable images to be transferred from front-end processors to mainframes for large database updates, graphics, simulations, and CAD/CAM/CAE operations. The results of these trials will be better requirement specs that correctly reflect, at the time of open request for quotes (RFQs), the appropriate needs of the data customers as identified in the trials, which will be published and defended in industry conferences and forums.

In this two-step approach, suppliers and providers can first explore new opportunities in trials, and then provide these services to the general public in a realistic and timely fashion. As we initiate low- and medium-speed services and explore the higher-speed opportunities, we are in actuality establishing a new U.S. customer base in the 1990s that will subsequently be ready to use the greater capabilities of new broadband network services, as they become available at the turn of the century.

Challenge six

Both private and public providers need to overlay value-added application services on top of the basic public transport services. As we consider numerous application opportunities, we see the need for platforms that provide additional work on the communication transport mechanisms or on information content. These centers can be shared across industries, as they deliver similar services for different applications. Both computer- and operator-assisted services can be provided for both voice and data messages, as well as image and video. For example, these new application service centers can provide data encryption, specialized routing, code conversions, protocol conversions, delayed delivery, broadcast, enhanced 911, 800 services, polling, sensing, alarm monitoring, number translations, transaction processing, data manipulation, image generation, videoconferencing, picture generation, file search and retrieval, data storage, operator database searches, and gateway services for a whole spectrum of common applications.

On the other hand, there will be a need for specialized CPE services tailored for specific applications, such as internal X-ray imaging programs for radiologists. Hence, there will be two levels of application centers that exist above the network. These service complexes need to be identified and programs launched with enough design lead time to ensure that the full range of offerings are available in a timely manner to stimulate new markets and continue to meet the expanding needs of new information users.

Challenge seven

Public network providers need to plan and establish new automated network-provisioning and support-systems services to enable rapid deployment of customer requests and dynamic operational, maintenance, and administrative control of the communications network. They need to establish a new communications support infrastructure that not only automates functions for network management in the traditional sense, but also ensures that the survivability objectives of the most stringent data requirements are achievable. Automated customer requests to provide dynamic bandwidth allocation, bandwidth on demand, alternate routing, and extensive error-rate management must be available for not only voice, but also data and video messages. As the world turns to communication instead of transportation, we cannot implement a complex interconnected private/public information network that cannot stand the test of security and survivability.

These support system complexes will begin to deliver many new operational and administrative services as new private network services are layered on top of the public offering. In this manner, future layered networks' layered services will be supported by an expanding infrastructure, creating a very competitive, feature-rich, interactive environment between user terminals and private and public networks. This will enable networks to become less restrictive and more transparent to their customers. It will then become the supporting infrastructure that is essential for sustaining new competitive, layered services from numerous providers in this blossoming, competitive information marketplace.

Challenge eight

We need to use a layered network's layered services model. As providers layer services to meet specific applications, the users will be accessing both private and public network services, as well as shared and specialized application service centers. This layering of network services can be effectively visualized on a layered network's layered services model as having identifiable entrances and exits that are clarified by standards as open network architecture access points. We need to establish this reference model. It could be viewed as one that goes from CPE internal networks (level 1) to network transport access systems close to the user for interconnecting private-to-public systems (level 2); on to the internal public network call processing, routing, and control systems (level 3); then to the application service centers, both shared and specialized (level 4); and finally to the internal CPE processing systems (level 5), which completes the work requested by the call. This type of structure enables work to be quickly localized to one level or another to clearly differentiate who does what, where, and how. This is essential if we are to avoid overlapping products providing overlapping services. Similarly, it is economically prudent to resist having a proliferation of application service centers performing similar functions or duplicating work performed by lower layers of the network (see Heldman, *Global Telecommunications*).

Challenge nine

We need new pricing tariffs and regulatory actions to establish the communication information infrastructure to support a competitive voice, data, and video marketplace. We need a structure in which the public and private offerings can coexist and grow. We must first take the time to understand the complexities of establishing integrated layered network services that enable interconnection of private and public offerings.

As the information marketplace demands more and more internetworking, interprocessing, and interservices for businesses, the home, government, and educators, there is a need for more ubiquitous offerings for both urban and rural customers. Hence, balance and scope are necessary. Application services cannot be limited by artificial restraints that make their achievement too complex, too expensive, or not attainable. As indicated in the "critical connections" report to Congress by its Congressional Office of Technology Assessment, regulatory understanding is needed in order for artificial boundaries to be erased. "We must obtain the correct infrastructure that ensures that a strategic assent of the United States is established. This is an ubiquitous public communications network that supports the complete spectrum of voice, data, and video information services to all the people in the United States, not just a selected few."

Challenge ten

We need to provide global information services. Our industries are not limited to the boundaries of the United States. Their offices exist throughout the world. As foreign mergers, partnerships, and acquisitions continue to occur, the boundaries and differences between foreign and U.S. firms are becoming a blur as the marketplace becomes more and more global. The next millennium will be the era of this form of economic expansion and interconnection. Value added networks (VANs) will lie over the globe, tying regions, countries, and continents together. As shifts take place from heavy-equipment, production-line industrialization to more software, information-based industries, locations for the workplace can become quite remote. Distance will no longer be a limiting or restraining barrier. This global information marketplace will require our participation in the form of interactive information exchange and use. Offerings provided within one region will need to be interconnected and available throughout the global information society.

Challenge eleven

We need to create a formidable, competitive work force. We cannot achieve the previous ten challenges without having knowledgeable, dedicated people. We need to educate our U.S. work force in order for them to participate in this exciting information marketplace. Customer requirements can no longer go unanswered due to resistance, ignorance, noninterest, or lack of understanding. We must embrace new technologies to offer new services. Broadband ISDN and narrowband ISDN technological possibilities need to be understood and applied to market opportunities. No longer does "Ma Bell" give her "baby bells" the selected product-of-the-time to operate. RBOCs and other local exchange carriers (LECs) and their suppliers must do their own thinking and planning. This means that long-term, five-year-lead-time products need to be identified and launched in order to be available in time for the marketplace. To achieve this, providers and suppliers need to use a planning process that supports preprogram, preproject, and preproduct participation by people from every aspect of the industry. The providers must formulate long-term planning, product-development relationships with the suppliers, who must work with their customers to identify, modify, and specify the desired features and services that meet the complete communication information needs of today and tomorrow.

The twelfth challenge

Providers and suppliers, both private and public, need long-term and short-term financial goals. The information marketplace will indeed become more and more complex, expensive, and competitive. The rewards will be great for the successful players. We need to establish a long-term relationship with our financial stockholders to enable us to establish the desired supportive infrastructure that will foster an ever-growing array of rewarding services. On the other hand, we must continue to meet the realistic expectations for today's return-on-investment objectives. We will need to turn to creative financial partnerships with a new breed of stockholders to achieve longer-term, higher-reward ventures in order to put in place both the shared and specialized layers of new information network services. (See Fig. 7-5.)

In conclusion

Here lie our challenges! The result is to create a new, growing, multilayered network service structure that enables internetworking, interprocessing, and interservices from multiple providers using multiple products from multiple suppliers to meet customers' expanding expectations for service applications. This is the challenge of the 1990s, the challenge of the information marketplace!

Strategies

So what do we do? We need specific strategies and plans for future marketing and technical endeavors. Therefore, the following overview describes a general program in terms of specific technologies that provide specific service offerings to specific users.

As noted earlier, to correctly position the United States and achieve a leadership role in this period of internal change, we need to have a detailed, clear vision denoting our key strategies for success. The industry has had a lot of activity during the past four years, but many sources have indicated that no one has stepped forward with a clear, innovative vision of where they are going and how they wish to get there.

A national plan of action is then needed to fulfill the vision. It should be based on key strategies and denote how we should play the game from technological, regulatory/nonregulatory, market, and management perspectives. Together, these underlying strategies and the plan of action should help us introduce the right type of networks, products, and services at the right time, taking into account the transition period required to encourage user acceptance and growth. As a starting point, let's review the strategies shown in Table 7-1.

ISDN services strategy (M)   Currently, "ISDN" is not considered "IS," but only "IDN." "IS" means "integrated services," and "IDN" means "integrated digital networks." We must begin to realize that ISDN really means "integrated networks' integrated services" (INIS). As we begin the long journey from simply providing ISDN access, we must develop a layered set of services that are supported by the OSI model to provide en-

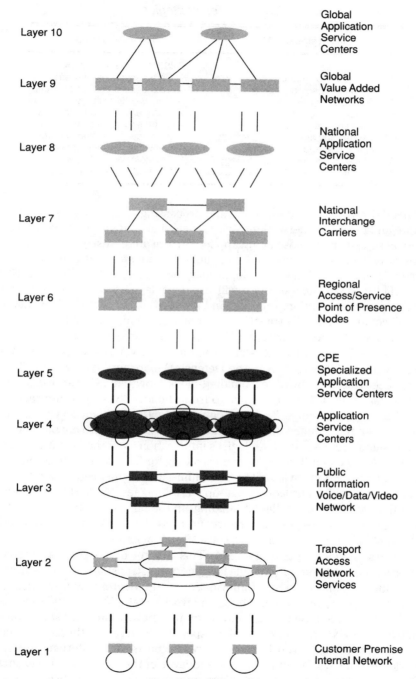

Layer 10	Global Application Service Centers
Layer 9	Global Value Added Networks
Layer 8	National Application Service Centers
Layer 7	National Interchange Carriers
Layer 6	Regional Access/Service Point of Presence Nodes
Layer 5	CPE Specialized Application Service Centers
Layer 4	Application Service Centers
Layer 3	Public Information Voice/Data/Video Network
Layer 2	Transport Access Network Services
Layer 1	Customer Premise Internal Network

Fig. 7-5. Layered networks' layered services.

Table 7-1. Ten strategies

Market strategy (M)	Technical strategy (T)
1. ISDN service strategy	—ISDN interface strategy
2. Data services strategy	—Data network overlay strategy
3. Information-pricing strategy	—Data trans/processing strategy
4. Information access strategy	—Gateway/IN strategy
5. Shared-services strategy	—Wideband transport strategy
6. Urban service strategy	—Metro switching strategy
7. Res/Bus/Gov applications strategy	—Information switching strategy
8. Service management strategy	—Network management strategy
9. Competitive/monopolistic strategy	—Private/public network strategy
10. Integrated network services strategy	—Layered networks strategy

hanced throughput, manipulation, and processing, as information requires more and more internetworking, interprocessing, and interservices. These services must track those being applied for private industry, as LANs and MANs become interconnected to enable file transfer and manipulation. Therefore, ISDN will mean basic, primary access, as well as H0, H11, H3, and H4 broadband access, but it will also mean layered protocols and code conversions, encryption, security, and reliability transport services. Later it will come to mean file transfer and presentation mechanisms.

ISDN interface strategy (T)   Note that data networks can be built without ISDN, now that 802 standards for LANs/MANs transport are defined without ISDN. But even the Defense Department requires an ISDN interface strategy to provide a building-block growth of ISDN access to the public facilities. This access must go from the low 2B+D 64K interface to the 1.2–50-gigabit range. This strategy must be for the home as well as the business so that a home 2B+D digital channel is available to enable a complete digital transfer of a radiologist's information from the hospital to the home. We need an ISDN deployment strategy for fiber to the home and the hospital, based on providing low-, medium-, and high-speed data rates to the home as well as broadcast dial-up HDTV movies. Similarly, intelligent buildings and not-so-intelligent buildings in the business community need to be interconnected, using narrowband (less than 128 Kbps), wideband (128K–45 Mbps), and broadband (50+ Mbps) facilities.

Data services strategy (M)   This market-based strategy recognizes the current computer needs for generating, transmitting, processing, manipulating, and presenting information to the doctors, lawyers, manufacturing, education (home, school, and work), state, federal, and defense users. "Data communications" and "computer networking" are sometimes called "data C&C networking." More specifically, AT&T's thrust is called "data networking." Even Bellcore's new jargon is now "information networking." This will become a major endeavor of the late 1990s as sensing (home incarceration), polling (alarm monitoring), and database access and information exchange become switched rather than point-to-point. Hence, a family of data-handling services needs to be available, services such as protocol conversion, code conversion, network retry, delayed delivery, text storage, record access, and file manipulation.

Data network overlay (T)   This is the data network overlay on the analog and digital voice networks. It enables a packet/circuit-switched data transport using the packet/circuit-handling capabilities of the switch suppliers to be deployed in a string overlay manner across cities. It must be provided ubiquitously in order to enable small businesses to use the network. Century 21, local drugstores, retail clothing stores, large banks, and hospitals share a common network. It is a "beginner network" strategy, which will be enhanced by later strategies as a "backbone level two and three network" for data offerings.

Information pricing strategy (M)   As noted in financial analyses on previous data packet pricing, 56K data pricing, T1 tariffs, etc., have not been used by the many—only the few. Public providers must reassess their pricing structure. Even usage-sensitive frame relay/SMDS pricing might not be the way to go as service providers attempt to monitor the "gigabit" throughput of information. Many studies have noted that fixed price levels with a range of throughput capabilities is probably the way to go, with limited or no access charges. This entry-level pricing strategy is designed to encourage growth and bring users to the more advanced services.

Data transmission/price strategy (T)   Today, 50% of a central office is point-to-point. As private networking encourages customers to leave point-to-point to switched facilities, LECs need to construct a transportation facility that has high-quality/availability objectives, especially once success occurs. This is the key to the data network world; it cannot go down. Similarly, its ubiquitousness is another essential aspect. As layers of protocols are available for file management, access, and transfer, the small business users will want to use the public network to interconnect their dissimilar systems. "Content" processing will be necessary as providers begin to provide "packaged" offerings to the home, business, etc. This strategy is to achieve an ubiquitous, high-quality data transport network that provides the full seven-layer Open Systems Interconnection (OSI) model of internetworking capabilities. It is designed to handle tremendous growth in high-volume traffic. This will require a shift from strategy two, the beginner network, to a growth network in sync with cheap transport pricing for both low-speed kilobits and high-speed gigabits.

Information access strategy (M)   As users internetwork information and access local and remote databases, they need a family of packaged data/image services that enable search, retrieval, manipulation, and presentation of information by interconnecting with gateways and intelligent nodes attached to the network. As more and more of the inquiry/response tasks of the 18 major industries become electronic, from patient record search to theater tickets/seating arrangements, data transport mechanisms can be shared across markets as they are packaged into specific application offerings. Note: the set of transport services are lower-layer OSI functions used by higher-layer OSI service functions, which are then packaged into specific offerings for user applications.

Gateway/intelligent network strategy (T) Many features are being provided through attachments to the network. Databases are available for 800 numbers, 900 offerings, shared features, and gateways accessing databases for videotex offerings. We need to combine these strategies into a set of layered gateway offerings as a building-block support structure. LECs do not need different mechanisms for accessing intelligent nodes using highly complex signaling access, while videotex dial-up services use direct connection access.

Shared services strategy (M) As many users seek to access the customer over "the last mile," we need to construct a list of offerings that will be provided by RBOCs and other LECs. These offerings could include: energy management, alarm monitoring, CATV, HDTV, patient information exchange, inventory control, library research, and common community information exchange. The result would be a set of common access/transport/manipulating/processing providers' services that could share wideband/broadband transport facilities to achieve closer and closer access to the user. As their services are interexchanged and internetworked, both the public and private sectors will benefit.

Wideband/broadband transport strategy (T) The access/local-loop facilities need to be restructured into multinode, variable-channel, higher-capacity, local transport loops that enable the full range of services, from low-speed data to high-speed, high-definition TV services that are available to both the home and small business communities. This has been described as a rings-on-rings strategy. It is the key to the growth of the public information network and should be reflected in the local plant topology and deployment plans.

Urban service strategy (M) In the urban service strategy, voice, data, and video transport and processing services are integrated into a family of service offerings with initial availability in the urban business community. This arena has a great potential for internetworking and inter-processing of multiple provider services. The ISDN international routing/addressing will facilitate private-to-public addressing, translation, access, speed billing, record keeping, and file management and will enable the higher levels of urban information exchange.

Metro switching strategy (T) The next-generation switch will be deployed as a central base to provide integrated voice, data, and video transport and control using fast packet and photonic switching capabilities. It will augment and replace the beginner start-up network of strategy one, which is also reapplied to the rural environment, time-phased for their delayed growth period. The metro switches' planning and development will take three to four years in order to be available in the late 1990s. Current digital voice systems will take on additional switching matrices until the new technology is available. Some suppliers say their current systems will simply grow and change shape, slowly discarding the original and adding a remote front-end access switch. Others believe in a new superswitch with a distributed switching structure. Whenever, whatever, however, eventually a new switch will be needed for new broadband services.

Residential/business/government application strategy (M) Residential markets need to define low-speed data network applications such as PC-PC, home to videotex, sensing (home incarceration), polling (patient monitoring, energy management), high-definition television, and shopping-at-home applications to denote low-speed/high-speed facility strategies for the home. Business, education, government, and state must do the same now that security is a new attribute in addition to speed and quality. They have become new issues, especially for law enforcement, banks, defense, etc. These strategies should indicate what services will be available on the specialized networks and what dependent strategies will require services from the other layers of the 10-layer information network service model (see *Global Telecommunications*). Finally, the interdependent service features must indicate cross-industry functionality. The results will show how the new Centrex, info switch, or feature switch can function with shared and private nodal-point systems, as switch modules are remotely located on customers' premises.

Information switching strategy (T) The next-generation information services switch will replace Centrex and compete with private information I-PBXs/IBXs for shared information services on a public network. It will address the private shared-tenant, home, small-business, and shopping-center environments. This switch is a key service-switch strategy. It is needed as a first- and second-generation offering as both N-ISDN and B-ISDN for baseband and broadband offerings.

Service management strategy (M) Who will be "the keepers of the network" as we internetwork many new private and public networks and interconnect many new services? Service management gives the customer, the private network manager, and the public provider access to network operations. It will be an essential network element as data transport networks are expanded and rerouted, depending on the traffic needs of the user and the time-of-day network operations. The set of network features and services that both private and government information managers want the LECs to provide are contained in the lower layers of the INS information network services (LNLS) model. RBOCs will also use them on a private basis in their higher application service center (ASC) layers. Note: the higher-layer services cannot exist without the lower network services. If these services are brought to the high level, they will be more expensive to provide ubiquitously.

Network management strategy (T)   A multilayered network management strategy must provide access and control for not only the public network voice infrastructure, but also for data and video. Also, the ability to administer and control networks will encourage the private networks to move from point-to-point to switched, shared facilities, especially as more private-to-private internetworking takes place through the public network. Hence, there is a need for a layered network management strategy between exchange, interexchange, special, and private carriers. Without it, we are simply building a forthcoming disaster. It is as simple as that!

Competitive/monopolistic strategy (M)   There might be a need for a shared local transport strategy to open up competitive "content" ventures by the RBOCs in the higher layers of the INS model. Then the individual providers can package their more specific application-oriented service packages to meet the full spectrum of user needs without having to form expensive partnerships, which increase the costs and make the offerings less attractive.

Private/public network strategy (T)   The internetworking of LANs, MANs, and WANs with the public networks is the key to long-term success, as noted in the INS model. A full range of standards needs to be established by the LECs that lead the way in this arena, especially the RBOC who "opens up the local monopoly." Note: this lower layer (shared ring transport with access switching nodes) is needed even if the shared option is not elected.

Integrated network services strategy (M)   As noted in the challenges, a multilayer service model needs to be developed to show how each of the higher layers will use services from the lower layer. Nested functions will be used by both the residential market and business community. They will reside at the lower level to be shared across the higher level. For example, gateway mechanisms will exist that are used by both residential and business customers to access videotex services. These will become part of each application package and will have different prices as more or less added value is deployed. (See the layered service model and the OSI seven-layer model in Heldman, *Global Telecommunications*).

Layered networks strategy (T)   The ten-layer integrated networks' integrated services model (Fig. 7-6), enables shared, private, advanced, and specialized networks to be deployed in both ubiquitous and selected environments, enabling multiple providers to communicate more effectively throughout the information marketplace. The information network model denotes new switching transport systems that meet the needs of the advanced-services and special-service markets. The info switch is used to provide residential/business nodal-point access to the advanced service market. The ring switch can be deployed to provide the base infra-

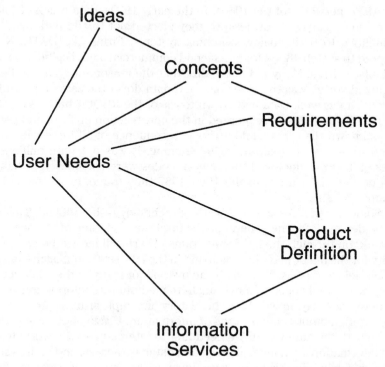

Fig. 7-6. Preprogram, preproject, preproduct planning.

structure for integrating private LANs, MANs, and WANs to the public network. Finally, note how communication service centers and network management centers can be deployed to meet the needs of all the layered networks. On this vehicle, the layers of information services can be delivered to user applications in both the private and public arena, across the residential/business/state/education/federal information marketplace.

Players should carefully consider these strategies and apply them to their areas of the marketplace. Where strategies do not exactly meet players' constraints or applications, they could modify, change, and enhance these strategies. However, it is essential to end up with a set of strategies that they all will see, understand, and implement. The strategies should clearly denote cross networking requirements and interdependence among market sectors and segments. (See Fig. 7-6.)

Perspective

If the United States does not establish public data networks, then what? So what? We have seen the military and the educational firms build their 465-L and ARPA networks of the 1960s. In the early 1970s, data was ½ of 1% of the common carriers' business, so they elected not to add data handling capabilities to their analog switching systems. Firms like DATRAN attempted to switch 4800-bits-per-second commercial data. Digital occurred in the late 1970s. Many did not realize that the reason for going to digital (digitized voice) was to move voice and data down the same pipe and to switch data as well. Data had grown to 3% of the RBOCs' business by the 1980s. Still its needs were ignored in the construction of T1 digital paths. It was known that data could arrive in a sequence of all fifteen 1s or 0s, which would cause a slip in digital framing. It was not until millions of miles of T1 were deployed that special codes were considered to handle data on digital facilities. In the 1986–1987 time frame, clear channel T1 became an issue.

Some say wait for the next wave of technology—broadband. This will not be deployed for the general public until the beginning of the new century and on through the 2020 time frame. There will indeed be an opportunity for the major plays in broadband in the late 1990s, but again this will be on a selected basis. This will be an overkill for many of the 26 data user types, who simply need to move small- to medium-rate information across the network today. In watching the history of computer-usage growth, the early users did not generate tons of information. Usage begat usage. So it will be with the data world. We need first to interconnect the world to enable information exchange, multisystem database access, and multisystem problem solving. In this manner we advance the computer evolution from large mainframes to PCs, to interconnected PCs, to PC to mainframe, to mainframe to mainframe, as we enter the world of distributed processing and proceed to intelligent processing and imaging.

So what will happen if the United States doesn't build the public data network? (See "Future of the Internet" in appendix A.) Well, data is only 3% of the business because it is going private. New alternative private networks will continue to be established. Switch manufacturers will move into the transport segment of the business if the current providers do not use their more data-oriented products. If not, someone will. By the late 1990s, the information age will blossom somewhere around the globe. Once successful there, it will be "cloned" here in the United States. This can and will occur in new urban information companies (local VANs).

What if the United States does go ahead and does immediately deploy ISDN narrowband and wideband public data networks? What happens? The answer is quite simple. Without needing to deploy fiber, by simply extending the capabilities of existing twisted pair wires, the U.S. carriers now have a new second service with an entirely new expanding group of customers, instead of trying to take an existing customer base that is growing at only 4–6% for voice lines. There is a new bottomless pit of potential data/information users. Once this infrastructure is established, then the marketing groups will be able to overlay application service centers, as the "information game" becomes one of services on top of services. As noted in one radiologists' trial, once radiologists transport X-rays at 64 Kbps or 128 Kbps, the real issue will become the moving of other entities besides their X-rays, such as patient files. Then it becomes an issue of storing X-rays and files, etc.

Using a layered house model, once the support structure is established, we will have a great deal of enjoyment designing and constructing the rooms on top and adapting them to our particular lifestyle and applications, be they bedrooms, bathrooms, offices, studies, lofts, or whatever. It should also be noted that narrowband offerings provide revenues to enable the LECs to more quickly deploy universal broadband fiber.

Thus we are indeed at a crossroads in time. This is a timely opportunity to play "the information game." For various reasons, the U.S. didn't in 1965 or 1985. If it doesn't step up to the plate now, perhaps the third time around the United States will be out. It might be time for another global player to play the information game . . . in another place . . . in another time.

Part 3

Plan of action

"*I would not sit waiting for some
vague tomorrow, nor for something
to happen . . . One could wait a lifetime,
and find nothing at the end of waiting . . .
I would begin here; I would make
something happen.*"

Louis L'Amour

8

Programs, projects, and services

*"Anything one man can imagine
Another man can make real."*

Jules Verne

The preceding analysis describes an overall plan for the future in terms of networks and provides specific service offerings to specific users. To correctly position the United States in a leadership role in this period of intense change, we need to have a detailed, clear plan denoting our key implementation actions for being successful. We need to step forward and define where we are going and how we want to get there. This will provide a vision for all to understand and follow.

A plan of action is then needed to fulfill this vision for the future. It should be based on the key strategies and denote how we should play the game from technological, regulatory/nonregulatory, market, and management perspectives. Together these underlying strategies and the plan of action should help us introduce the right types of networks, products, and services, at the right time, taking into account the transition period required to encourage user acceptance and growth.

More detailed strategies within the plan of action need to transition from the strategic to the tactical. For example, we want to win the war. To win the war, we need to take three islands. We will use the Marines to take island one. We will blockade island two with Navy ships. We will drop bombs and paratroopers on island three. The Marines will hit the beaches at 0600 on the 31st. The Navy will begin its blockade by noon on the 31st.

The 101st Airborne will drop at 0900 after three hours of bombing on the 31st. Let's get on with it . . .

Likewise, it is essential to play the information game based on layers of strategic marketing and technical strategies. These MARC-TEC (market-technology) strategies are tied to business strategies, which then formulate the basis for financial strategies. These are then in concurrence or cause change in governing-policy strategies. It doesn't matter whether we move from governing policy to the market or from market to governance. What is important is that one can change the other, as the analysis recursively cycles through these levels until "where we want to go" is established from all three—marketing, technical, and management—perspectives!

These MARC-TEC strategies will be structured onto a multilayer network services model that is designed to enable both regulated and nonregulated services to blossom in an interrelated private/public marketplace. Hence, one of the key outputs of the analysis will be to understand this structure and then cross-relate present as well as future projects onto this model in order to appreciate how they must interrelate together. To achieve this, we need to formulate a comfortable vision of the future based on specific MARC-TEC strategies that map onto competitive directives. One such directive might be to become market-based firms providing cost-effective services that produce growing revenue from specific market applications, leveraging the existing network's capabilities to enable information network and service providers and suppliers to achieve a satisfactory return on investment for their owners.

Introduction

The billboard across from the airport said, "Data Message Leaving To New York Every Fifteen Seconds." What a pity that it didn't also say, "Videoconference Leaving to Anywhere in the World Every Fifteen Minutes for $1.00 Per Minute." Current video communication centers' costs range from $300–$500 per hour using satellite. Two T3s from Minneapolis-Omaha cost around $6000 per month or $10 per hour or 16¢ per minute on a point-to-point basis. By the late 1990s, $1 per minute could be reasonable for switched videophone on an usage basis, using 3 megabit (2 T1s or DS1s) capabilities of a gigabit fiber, perhaps even 2 T3s or SONET 155 million bps by the year 2010.

It is becoming harder and harder to be in two places at the same time, let alone two distant places. They say distance makes the heart grow fonder, but as we commute through growing transportation problems, it might be safe to say, "Distance makes the heart beat faster."

So what about the slogan, "Communication Instead of Transportation?" How can we use communication to make our life a little easier? A Washington newspaper has a special column called "Bumper to Bumper."

One article consisted of a farewell letter from a couple who decided to quit their city jobs to get rid of their 5-hour (2½ hours each way) commute. They were both highly capable in their work, but they were incapable of handling bumper stress any longer. They were heading off to Wyoming—some place west of Washington D.C.

So what can a telephone company do to promote growth in its more rural environments? How can it help eliminate the stress of customers in the cities? How can it increase the productivity of its companies, governments, educational facilities, protection agencies, retail stores, medical facilities, and small businesses in its operating region?

Time is important. Using better communications, users should be able to access remote information easily and instantaneously. They must be able to feel secure that our nation's communication networks will always be there—independent of "information congestion." They must be able to survive a network disaster. Anyone who was in the 1988 Hinsdale, Illinois disaster area felt total frustration and isolation for not just a few hours or days, but for weeks. Imagine if the hospitals had gone fully computer automated using switched facilities, not point-to-point. Imagine if MCI and Sprint had not come into the area from different paths. The disaster would have totally crippled the entire community. As it was, it was painful enough.

No, a new network cannot be built that handles voice, data, and video traffic without structuring it to handle both success and failure. We should not spend billions on an architecture that causes high risk to its operation if usage increases. We want to build a network that is designed to foster more and more usage, as it provides more and more features and services that help us live better, less-stressful, more-effective lives.

This means we must provide a network that enables more and more networks to interconnect together. They say that a computer's capabilities increase is directly proportional to the number of computers to which it is interconnected. Similarly, a small-town doctor's effectiveness is enhanced proportionately to the number of specialists that can be directly reached for consultation without physical travel.

We also know that inventory control and management are essential to the spare-parts industry. Here again, data networks are the key to their economic success. Sadly, we have seen the large universities' desire to extend their supercomputer capabilities to small universities in the area go unanswered. We have seen the radiologists' desire for high quality, faster information transfer to their homes go unanswered. (Groups of these specialists spend an average of 150 hours every weekend going to and from the hospital.) We see urban sprawl spreading with its own pollution and congestion, while the population in the rural United States is simply disappearing. Only a very small percentage of our citizens now remain in our more remote communities. As we look at applications for various users' communities of interest, we see the need to tie together all aspects of their daily operation. This cuts across all business units, such as small business doctors and pharmacists who need to communicate with large business hospitals, poison control centers, 911, etc. As Figs. 8-1 and 8-2 so note, we must fully interconnect all aspects of the entire community of interest.

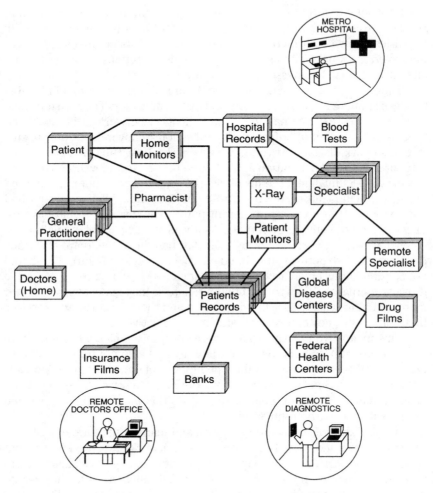

Fig. 8-1. Medical community of interest.

So what can providers do to capitalize on these conditions and events and turn them into a successful opportunity for both themselves and their customers? It is time to think bigger than the voice telephone network. In looking at the new sets of services that could be provided, one cannot help but notice how predominantly voice-oriented they currently are: voice

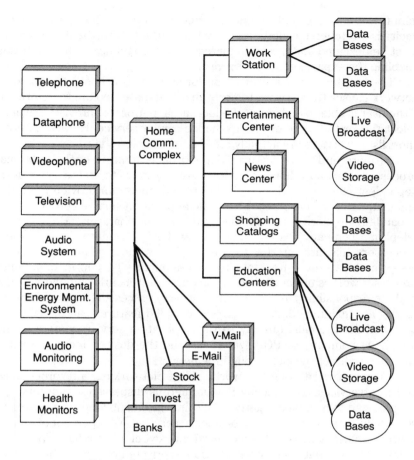

Fig. 8-2. Home communications center.

mail, calling-party ID, specialized voice-call transfers, distinctive ringing, and specialized billing. Customers are offered more network control of their voice calls, etc., etc. Yes, these are all good, but we must be aware that we shouldn't try to provide voice-only solutions.

These are new needs that must be addressed. Our customers now require a new, complete, ubiquitous, inexpensive, highly secure, and serviceable data network. This network should offer a whole host of new features and services to meet the needs of many new types of data customers. These data customers will expect the network to be transparent to them, at least to the same extent that the voice network is today.

However, the voice network was really priced to limit growth and usage: Remember how fast we talked on personal "long-distance" calls? It is better today, but still very expensive (even at U.S. Sprint's 10 cents per minute). Premature central office replacement is still a concern as the capacity is stretched by suburban growth and usage. Now the data network must be designed and priced to increase growth and usage. Key to this is enabling the expansion of new services.

The traditional voice user's "call for service" (off hook) to the voice network causes the network to perform a "database" lookup to determine if the user is able to dial out of the local area, perform abbreviated dial, or select this or that type of interexchange carrier service. However, only a short list of services are actually offered to the user today.

As more and more features depend on user preference, additional processors can and will be added to existing systems to access a new user-dependent class of services. Telephone companies can provide more dynamic updates to these tables from the application service centers, using expanded-supplier, recent-change database mechanisms. Hence, user telephone company feature changes can be implemented in additional processors that are internal to the supplier without requiring software hooks and external interrupts to programming. This form of adjunct processor work is under the direct software control of the suppliers, and it is their responsibility to ensure that access, processing, and system integrity are maintained. Additional value-added features can be achieved by switching the calls directly to the application service centers in the telephone company or POP, which can add the desired increased value and then redirect the call through the network.

We are now entering the world of internetworking, interprocessing, and interservices. This means that numerous dispersed computers and people need to be linked together. This linkage can no longer be limited to voice only, with everything else on special services. Fifty percent of today's central offices consist of specialized nonswitched services (point-to-point). We can no longer foster isolation of private networks. Their proliferation should be channeled to interconnect to public networks, or we will return to the days of the late 1920s when none of the newer telephone companies could talk to each other.

Today, talk means total communication. If this is so, then the U.S. firms have an opportunity to recognize this mission to be the "world leaders" in the movement, handling, and processing of information. With current MFJ restrictions, RBOCs are indeed severely limited in the processing arena. But this can also become an asset because it ensures that RBOCs provide a vehicle for anyone and everyone to provide data manipulation offerings to U.S. customers. Third-party services and joint ventures in the private arena can be established as the architecture encourages more and more private-to-public-to-private-to-public information movement and exchange.

New services must be available to enable the whole person to communicate. This means not only the ear, but also the eye. The eye can see data displayed in table form or in translated images that enable the mind to more quickly grasp the data's meaning and ramifications. The logical extension of data is high-quality graphics and scenes in the form of slides and pictures. Finally, the ultimate is to see scenes, images, and people in high-quality color in a manner that has no distortion caused by rapid movement or change of information.

Needs change. This has been dramatically demonstrated in the past by comparing new information opportunities to past ventures. We saw the car change from simply a vehicle to go from point A to point B to a vehicle that provided comfort and entertainment.

In the early 1920s, a Sunday ride also meant standing in the rain to obtain button-down curtains from under the seat or getting out to remove a spare tire from under the rumble seat. There were usually one or two flats each Sunday on a short family ride from Yonkers, New York to the beaches of Fairfield, Connecticut.

Today a car must provide for many human comforts other than raw basic transportation. I remember obtaining a ride from a young fireball one night when my car broke down. He had a huge engine on a steel frame with a wooden bench for a seat. It was a little scary looking down as the road went past at speeds over 60 miles an hour.

Cars today attempt to offer (to one degree or another) heating, cooling, stereo, power steering, bucket seats, security, and survivability. So it will be with changing communication features and services. Large business network managers first went to private networks to help control costs, but as more capacity became available, they went from networks that offered 1,544,000 bits per second (T1) to those providing 45-million-bits-per-second (T3) capability. Now they are finding they can better use communication controllers that offer users many new service features, so they have moved from being network managers to becoming service providers.

So it must be with public carriers. They must provide a structure that enables "layers of networks" to provide "layers of services." In this manner, a single application can be achieved by obtaining many of its services from different networks and service centers until the application solution is achieved. A reference model is needed to identify where services are provided. A ten-layer model has been recommended. Its first layer (layer 1) provides for customers' internal communications and external entrance and exit. It then provides for network access at points as close to the customer as possible. At this second layer (layer 2) there might be some sharing of the public transport. It could be accomplished using superhighways that ring metro cities and extend to college campuses and shopping centers. Exit can be accomplished at ring switches to POPs (points of presence) of other providers. Hence, the local monopoly does not become a bottleneck, especially since anyone can drop direct wires or send radio signals to the user from these access points. (It might be necessary to share ownership of these rings to obtain justice concurrence so the bottleneck to the user is substantially reduced. This should then enable RBOCs to play with content and provide inter-LATA interstate networks. Otherwise, Congress will have to further resolve the continuously changing array of feature-dependent ONA issues. See "Critical Connections," Congressional Office of Technology Assessment Report to Congress on Local Monopoly in a Competitive Arena, July 1989, New Telecommunications Act Bills, 1995.)

To obtain network routing, address translations, class mark services, etc., in the public arena, the call can then be moved to the next layer of the structure, which has not only the traditional voice switches and their support systems, but also will include the data and video switches and their support systems. Above this layer (layer 3), value-added service can be provided by having several layers of service centers that provide for shared and specialized public and private offerings.

The fourth layer (layer 4) of the model consists of application service centers where common gateways, information centrexes, and data operator assistance centers exist. They provide advanced services over the manual transport offerings of the lower layers. Finally, the top layer (layer 5) of the model consists of specialized application service centers that offer unique features and access to specific users. A specialized poison center (layer 5) provides information to the lower-level 911 application service center (layer 4) that called it, or an internal hospital system on customer premises (layer 5) provides access to a patient's file in its patient history database archives.

Both the top and bottom of this local five-layer structure exists on customer premises, where at the lowest layer of the model, internal switching and input/output access are pertinent, and at the top, internal processing is achieved. Tying these two together completes the local model, such that:

- Layer 1. Specialized customer-premises internal networks.
- Layer 2. Network transport access is achieved.
- Layer 3. Network call-handling, routing, translating, billing, maintenance, and control exist.
- Layer 4. Advanced application service centers provide gateways to databases or where value-added presentations and data conversions are performed.
- Layer 5. Specialized customer-premises services where specialized database access and computer processing are provided.

In using this type of structure, we can readily map proposed solutions to meet customer needs and differentiate where and how private and public service offerings should be provided and interconnected. In this manner we will be able to use services from shared platforms to obtain economies of scale, as well as obtain unique, customized services for markets willing to pay the price.

Secondly, we can promote as much use as possible from the common bases to support more ubiquitous deployment. Hence, the higher layers (layers 6–10) benefit by having more users able to access their services. However, it is essential that their access and use do not in any manner impair the operation of the lower supporting structure.

To achieve this structure, we need to take both a technical and marketing point of view to better integrate strategies in order to achieve more successful offerings. We do not need to just deploy an ISDN strategy.

So it has been done in the past with previous data offerings and other services. To be successful, services must all be linked together in a continuous chain of strategies, from ISDN to data networks to data processing to wideband switches to network management to application service centers, as well as to private-to-public internetworking. All should be based on the layered networks' layered services model to show where, when, and how new networks, products, and services should be deployed so we know where one leaves off and another takes over. This linkage of strategies is called, for lack of a better name, the MARC-TEC (marketing-technology) strategies. (See Fig. 8-3.)

MARK-TEC strategies

The marketing/technical arenas for ISDN, data overlay, data processing, private/public interconnect, wideband/broadband switching, application service centers (gateways, information centers, specialized services), network management, support systems, layered networks, and layered services need to be addressed as a package for both today's and tomorrow's needs. We must not be too nearsighted or shortsighted, where only short-term planning strategies prevail. All the interdependent ramifications need to be visualized across these entities so that every opportunity is exploited to provide new revenues for achieving the long-term infrastructure (see Table 8-1 and Figs. 8-4 and 8-5).

Recommendations

With these considerations in mind, the following recommendations, similar to the challenges, should help U.S. providers and suppliers achieve an exciting array of successful new service offerings as they continue to upgrade the U.S. infrastructure.

Overlay digital network

Telco providers need to continue deployment of base-satellite digital switching systems in the rural arena. However, we should only obtain systems on a cluster basis using remote switch units. The purpose of this system complex is to put the base in the county seat and take out all the small

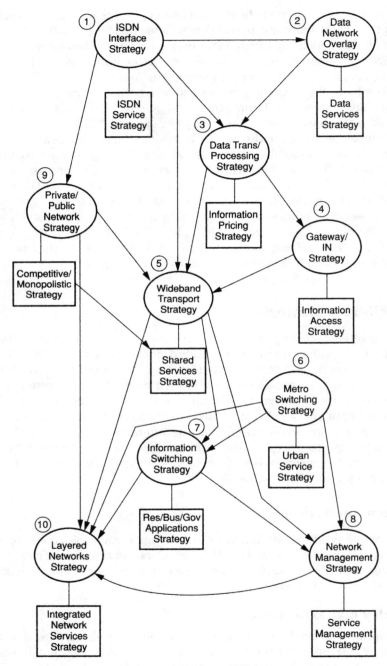

Fig. 8-3. MARK-TEC strategies flowchart.

Table 8-1. MARK-TEC strategies

Strategy 1: ISDN

ISDN interface strategy
- ISDN standard connector
- ISDN basic rate connection
  - —2 B channels (64K)
  - —1 D channel (16K)
  - —4 wire (4 kilofeet)
  - —192K b/s (144K b/s)
- Non-ISDN to ISDN converter
- Network terminating equipment two (NT2)
  - —Control of up to 8 ISDN terminals
  - —Address control/message identification
- Network terminating equipment one (NT1)
  - —2 wire to central office
  - —160 kilobits/second (15 kilometers)
  - —Special coding techniques
- ISDN primary rate interface
  - —23 B channels (64K)
  - —1 D channel (64K)
  - —1.544 M b/s
- ISDN H0 rate
- ISDN H11 rate
- ISDN H3 rate
- ISDN H4 rate
- ISDN/Syntran/SONET rates
  - —Syntran
  - —SONET
  - —SXS rates
  - —OC-X rates
- ISDN first three layers (OSI model) network control
- ISDN/IEEE 802.6 network interfaces
- ISDN home communication bus interface
- ISDN rate adapters
- Inverse/reverse multiplexers
- Inexpensive interface chip set

ISDN services strategies
- Pricing strategy for ISDN rates
  - —Basic
    - ~T interface
    - ~U interface
  - —Primary
  - —Broadband (greater than 1.544M b/s)
- Non-ISDN-ISDN interface rates
- Multiple ISDN common bus terminal controllers
- "D" channel services
  - —Alarm monitoring
  - —Remote environmental control
  - —Class interface services
- Secure/encrypted code control
- Audit trail
- Sensing interfaces
- Polling interfaces
- Videophone (128K bps)

Table 8-1. Continued.

Strategy 2: Data network services

Data network overlay strategy
- Collocate data switches/central office
- Remote data switch units/multiplexers/pads
- Nodal access data channel controllers (class 6)
- Nodal access level data network managers (class 6)
- Public/private access node ring interfaces (class 6)
- Data switches for automating internal operations
- Data switching/transport error rate control
  —Circuit
  —Packet
- ISDN
  —B channel data interfaces/control mechanisms
  —D channel data interfaces/control mechanisms
- Access to data network interexchange carriers (class 6)
- Delayed delivery/store & forward mechanisms
- Protocol converter/code converter mechanisms
  —Class 5, 6, 7
- Service communication center
  —Information centers/information switches
  —Metro switch interfaces strategy for data
  —Ring switch interfaces for data

Data services strategy
- Local area data network access
  —Packet
  —Circuit
- Data interexchange mechanisms
  —Protocol conversion
  —Code conversion
  —Async-sync
  —Delayed delivery
- Polling services
- Sensing services
- Broadcast services
- Error control services
- Security services
- Applications
  —Home incarceration
  —Remote patient home care
  —Inventory control
  —Status management
  —Inquiry/response services
  —Transaction services
  —Remote documentation
  —Data mail
  —Videophone

Strategy 3: Information transport services

Transport strategy
- Point-to-point networks
  —H0, T1, T3, T5, SXS, OC-X
  —1.2K b/s, 2.4K b/s, 4.8K b/s, 9.6K b/s, 19.2K b/s, 56K b/s, 64K b/s
  —1.5M b/s, 6.3M b/s, 45M b/s, 51M b/s, 155M b/s, 600M b/s, 2.45 b/s . . .
- Ring switches
- Metro switch
- Network management

- Specialized data interexchange carriers transport codes/mechanisms
- Computer/terminal interface units
- Error rate control mechanisms
- Alternate route
- Spare span switching mechanisms

Information movement service strategy
- Pricing
  - —56K
  - —T1
  - —T3
  - —Frame relay
  - —SMDS
  - —Syntran
  - —SONET
- Alternate routing
- Dynamic bandwidth allocation
- Customer bandwidth precall control
- Customer network management control
- Dedicated network management
- High-speed fax
- Video conferences
- Image 1

Strategy 4: Information network access strategies

Information gateway/intelligent network strategy
- Service communication centers
  - —Access to databases
  - —Image filing/management
  - —Delayed delivery
  - —File building
  - —Addressing
  - —Broadcast capabilities
  - —Off-network signaling
  - —Calling number identification
  - —Password/key identification
  - —Billing setup
  - —Call/packet duration monitoring
  - —Operator data call assistance
  - —Multiple database search
  - —Computer-to-computer interface
  - —Computer-to-terminal protocol conversion
  - —Data manipulation/presentation
  - —Videophone gateway access

Information access strategy
- Data access services
- Record buildup services
- Multiple database searches
- Operator assist services
- Artificial intelligence automated searches
- Data manipulation graphic representation
- Image buildup and control
- Data presentations
- Database updates
- Library search programs
- Data utilities—Seek and destroy
- Protection services
- Confidentiality services
- Real-time services

Table 8-1. Continued.
—Interactive inquiry/response
- Offline services
  —File manipulation
  —Graphic presentation
- Videophone access services

Strategy 5: Wideband

Wideband transport strategy
- Fiber deployment topology
  —Business
  —Home
- Ring switches
- Alternate routing
- Switches/nonswitched channels
- Twisted-pair/fiber strategy
- Internetworking gigabit information
- HDTV strategy
- Picturephone III strategy
- Videoconference services strategy
- Class 6–class 7 interfaces
- Rural switch access

Shared service strategy
- Cable/RBOC networks
- Multiple provider rings
- Shared bypass
- Customer network management
- Point of presence (POP) access
- Program distribution systems (third party)
- Class 6–class 7 interfaces services
- Shared transport rings
- Shared service pricing strategies

Strategy 6: Urban

Metro switching strategy
- Class 5 voice/data video switch
- Public ISDN to private ISDN routing/control
- ISDN numbering system
- Access to test/maintenance systems
- Access to automated service order centers
- Access to customer network control centers
- Access to service communication centers
  —Information switch
  —Feature switch
  —Intelligent nodes
  —Gateway nodes
- Interface to ring switch
- Route control to interexchange carriers
  —Voice
  —Data
  —Video
- Access to value-added networks
- Access to global VANs
- Access to value-added resalers
- Access to third-party programs
- Rural switch access

Strategy 7: Switching applications

Information switching strategy
- Distributed switching structure
- Standard interfaces from remote to base using ISDN "D" channel
- Remote communication centers (class 7)
  - —Home
  - —Residence
  - —Multiple tenant
  - —Business
    - ~Small
    - ~Large
  - —Education
  - —Government
- Remote service units (nonclass)
  - —User databases/program access
  - —Internal RBOC database/programs
  - —Centralized maintenance/control centers
- Voice, data, video information transfer
  - —Packet
  - —Fast packet
  - —Burst
  - —Circuit
- Security/error control/audit trails
- Billing data collection

Information application strategies
- Service communication centers
  - —Voice/data/video
    - ~Inquiry/response
    - ~Transaction
    - ~Data collection
    - ~Data distribution
    - ~Remote documentation
- Information center application programs
- PC-PC network information interexchange
- Internetworking private networks
- Interprocessing control programs
- Nested/layered program controllers
- Customer remote communication controllers (class 7)
- Polling/broadcast
- Specialized network management
- Specific application programs
  - —Hospitals
  - —Lawyers
  - —Police
  - —911
- Community-of-interest interconnection
- Videophone—videoconference services

Strategy 8: Management

Network management strategies
- Control of layered networks
- Internetworking managers
- Channel managers
- Disaster recovery managers
- Multilayer network management strategy
- Global network management interfaces
  - —Voice/data/video
  - —IECs, VANs, etc.

Table 8-1. Continued.

- Alternate routing ring control mechanisms
- Code block mechanisms
- Privacy control mechanisms
- Security control mechanisms
- LANs-MANs interface mechanisms

Service management strategy
- Layered customer network access/control
- Access to higher services of OSI model
- Access to operator assist network management center
- Access to multiple provider network management centers
- Rerouting and message delivery services
- Survivable network services
- Restorable (1–24 hour) network services
- Controllable network services

Strategy 9: Private/public

Private/public network strategy
- Internetworking
  —Routing/addressing/billing
  —Private—public
  —Public—public
  —Private—public—private
  —Non-ISDN—ISDN
  —SNA—DECNET—SAA
  —ISDN—IEEE
  —LANs—MANs
- Interprocessing
  —Remote/local database access/update
  —Distributed program applications
    ~Trans computer networks
- Interservices
  —Trans RBOCs
  —800
  —911
  —CLASS
  —Multiple database searches
  —LANs/MANs—bridges/gateways
  —Addressing/routing
- Access nodes (class 6)

Competitive/monopolistic strategy
- Shared ring transport
- Integrated RBOC services
- Integrated provider services
  —Transport
  —Databases
- "Content" regulatory strategy
  —Local monopoly
  —Competitive marketplace
  —Access rings
  —Class 6, class 7
  —Service communication centers

Strategy 10: Layered

Layered network strategy
- Multilayer model
- Class 7 (CPE switch)
  —Remote communication centers

- Class 6 (ring switch)
- Class 5 (metro switch)
- Service communication center
  —Gateways
  —Information switch
  —Remote service units
  —Intelligent nodes
- POP
- LANs/MANs/WANs
- Multilayered network management
- Private/public interconnectivity
- Global internetworking

Layered services strategy
- Layered model
- Specialized services
- Addressing services
- Public services
- Shared access/transport services
- Service communication centers
  —Information switches
  —Features switches
  —Third-party software
- Class 7
  —Network communication controller
  —Service communication controller
  —Information I-PBX/IBX

---

towns in the area with satellite remote switch units, thereby reducing NXX office codes and linking them all together with digital fiber and digital radio links. In this manner, more automated network maintenance and administration are achieved. The error rate of the network is changed to 1 error in 10^7 bits per second, upgraded from 10^3 or 10^4. Data packet switching is available for data calls as ISDN interfaces are provided to the user. (The ISDN interface for basic rate is two 64-Kbps channels and a 16-Kbps signaling channel. For the primary rate, it is 23 64-Kbps channels and a 64-Kbps signaling channel. The latter is usually for larger businesses having PBXs, etc.) Similarly, the urban/metro/suburban applicants can be addressed by overlaying these remote satellite base systems on existing offices, collocating a base where necessary to extend digital throughout the community, allowing ISDN interfacing to the local businesses, offsetting growth, reducing the number of suburban offices, and consolidating NXX office codes.

Overlay data network

Telecommunications network providers should provide an overlay data network across the entire area to provide cheap, low- to medium-speed data handling. Data users are not only transaction user types but also inquiry/response, data collection, data distribution, and remote documentation types. Inquiry/response type data users are indeed as big or bigger

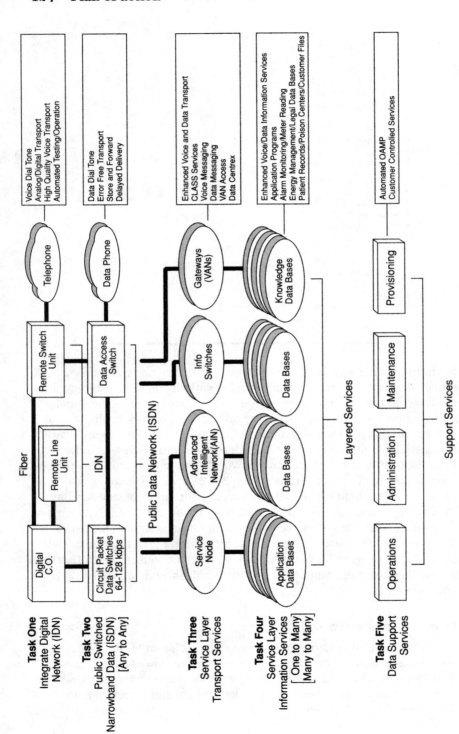

Fig. 8-4. Narrowband plan of action.

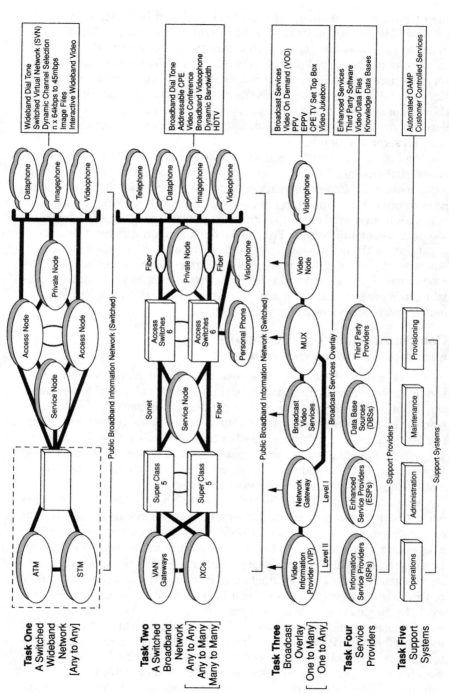

Fig. 8-5. Wideband/broadband plan of action.

than transactions, as Judge Greene once noted in his videotex network proposal. Hence, not only do we need computer-to-computer, but also terminal-to-computer and terminal-to-terminal interconnectivity.

The network should provide a whole host of new shared and unique services that are transparent to the users. In this manner, layers of services can be provided on top of the more basic public data offerings, which should consist of: packet, transport, network access, network blocking, network routing, three attempt unit, error-rate control, alternative routing, and sequence delivery.

High layers can provide polling, broadcast, protocol conversions, code conversions, delayed delivery, encryption, list assembly, message filing, database access, data manipulation, and data presentation features. These layers would allow supercomputer centers to provide specific applications packages from shared computers to such applications as the small business tool-and-die manufacturers. This will enable the many different computers to interact with the network to communicate with their supercomputers. Similarly, the many different computers in the doctors' office will need to reach unique pharmaceutical minicomputers as well as the various hospital mainframe computer systems. Of course, these data interface services must provide for increasing growth and usage.

Communications service centers and application service centers

As noted earlier, there should be several layers of service centers. One should provide advanced services for the network, such as videotex gateways that enable users to know what databases are connected to the network or operator-assist centers for data access. (Some people will simply request help to access databases rather than remember access keys, passwords, etc.) Again, polling and broadcast data capabilities can enable other providers to provide new services such as alarm monitoring, energy management, and environmental control systems.

Voice services, such as 800/900 database access, can be provided above the network using #7 signaling systems to communicate to and from existing systems. Alternatively, CLASS-type features for customer-line, side-access services can be added to existing systems using an adjacent processor (applications processor) to interface directly to existing systems' software. However, products that interrupt the call-processing flow of the existing network systems can have disastrous effects on the network. Strategies that change internal databases or "switch the call to the work" cause minimal interruption to the network. If additional work needs to be performed on the line side that cannot be achieved at the switched-through workstation, a new module needs to be designed in the existing switch using internal processors or software modules. Alternatively, users can "home" directly onto the specialized or advanced feature workstations.

Typical work for application service centers will be: delayed delivery mechanisms of data messages for datagram-type offerings, priority messages, barred access, messaging, record/file (management, access, retrieval, and storage), distribution of "online" information in real time from stock exchanges and news bureaus, and search-and-retrieve mechanisms to cause searches of distant or distributed databases (for example, missing persons searches through telephone listings).

On the other hand, specialized systems are needed to enable businesses to communicate with home shoppers to provide local sale announcements, achieve point-of-sale transactions, enable debt card transactions at local gasoline pumps, achieve the selection of airline reservations or seats at the local theater, and establish dinner reservations. These types of information center services will be provided by second- or third-party software programs located within the centers' databases. The somewhat unique offerings can be located by graphic menus provided at advanced application centers.

If extending the call continues to be prohibited by the MFJ across RBOC's LATAs, then other providers will perform this service through their specialized centers. In any event, the service will be provided by someone's application service center—either public or private, common or specialized. Note: We might want to call the communications public service centers "communications service centers or service nodes" (layer 4), then specialized, unique, private centers (layer 5) could then be called "application service centers" to better differentiate the level and type of work.

Once information providers (IPs) own (or share) application centers, pride of ownership will prevail as they then begin to fill the centers with an ever-increasing array of services. These centers will also have network access to remote modules located on customer premises. They should use standard interfaces or direct access options to reach their residential customers and small business communities. For example, an information center system complex can be centrally located with remote modules for residential, shared-tenant service (information switch with physically remote service modules).

Overlay wideband/broadband network

Medium- to high-speed data communication needs of up to $N \times 64$-Kbps of T1 to T3 rates will be met by overlay wideband data networks for 85% to 90% of the users over the late 1990s time span. But the remaining 10% to 15% will need a broadband system that switches T3+ rates of 45 million bits per second to 600 million bits per second using 2+ billion bits-per-second transport rates. Before constructing this complex system, limited broadbased test trials should be planned and implemented in metro communities, such as the Minneapolis COMPASS trial (see appendix A). There, 3M's MAN, Control Data's Ethernet, University of Minnesota's supercomputers, radiologists' X-ray systems, Creative Visual's Imaging systems, CAD/CAM designers, and video picturephone users could share the same switched network, as they exchanged information within their own unique community of interest. Here we have service trials that are not simply network trials. For example, in the Japanese "Metica" trial, they found the need for low-speed data as well as broadband data information. Doctors wished to not only see their remote patients, but also to obtain sensory data from them that indicated their temperatures, heartbeat, etc.

Subsequently, an overlay broadband ISDN network can be constructed that interconnects with LANs providing switched WAN and MAN capabilities to interface to systems such as IBM's SNA, using CCITT's Hx series ISDN interconnects as well as Bellcore's SONET (50 Mbps) rates. This will enable providers to more economically deliver switched megabit capacity to the remaining 10% of the information business customers requiring this higher-speed capability over narrowband and wideband.

Integrated ISDN network

We must begin planning now to replace the overlay networks by the late 1990s through the turn of the century in the metro environment. The overlay systems can then be moved deeper within the more rural communities to extend these systems' life and promote data growth and usage. There is considerable concern that the internal information flow will bring with it an information explosion, which will bury itself in its own growth. (Note: the occasional shortage of fax paper as fax-mania spreads across the United States today.) If priced right, these should bring a great deal of success. This success will encourage more growth, which will require a new set of more integrated voice/data/image/video services. This will also generate more traffic than an overlay network can adequately handle. Hence, a new generation of switching, processing, and support centers will be required. The multilayers of networking will need new network management centers for each layer. New standards will enable the higher level of computer information exchange as indicated by the CCITT Open Systems Interconnection (OSI) model's fourth through seventh layers. International internetworking standards for gigabit movement of information will be mandatory. Videophone to the home will still be the challenge of the late 1990s. A new, high-definition video system test trial for the home will be needed in the mid-1990 time frame. Photonic switching technology, VLSI design, and artificially intelligent network management systems will use layered software techniques, enabling more integrated but more specialized switching fabrics to better handle intermixed, low-speed voice and data with the video and broadband data bit streams.

This will require new products for each layer of the network to provide better integration and access to these huge information pipes as they are located closer and closer to the users' CPE systems. These could include: LANs, MANs, I-PBXs/IBXs, communication controllers, personal computers, videoconference centers, alarm systems, supercomputers, video-phones, or telephones. In looking at this evolving state, there is a need to quickly formulate not only a visionary structure, but also the products needed to achieve its reality. These products will take three to four years before becoming available in enough quantities to formulate the new network. They should consist of:

- A new fiber deployment topology to distribute wideband capabilities closer to the users using ring, star, and bus structures.
- CPE switches to provide initial services and switching as well as input/output using ISDN interfaces for narrowband (basic), wideband (primary), and broadband rates via a new family of D-channel signaling systems services that interface network D-channel controllers and #7 (or #8) signaling systems (class 8—internal switch modules).
- Network transport access system to provide access, handshaking control, and transport control mechanisms for voice, data, and video information. These new, distributed, automated-mainframe channel switches with extensive network management will identify information content, as well as provide S&S (secure and survivable) information transfer as class 6—ring switches.
- Metropolitan superswitch information controllers to provide integrated routing, addressing, non-ISDN/ISDN switching for simultaneous or separate voice, data, and video offerings. The controller handles movements of 100,000,000+ ATM packets of information using extensive data error rate coding mechanisms, survivability programs, audit trails, alternate route strategies, and extensive variable-bandwidth (channels) of STM CBR traffic for 50 Mbps videophone-type conversations. The next replacement for the class 5 offices will be initially within the metro community for voice, data, and video offerings. A metropolitan system, using next-generation switching techniques, sophisticated support centers, and network managers, will enable extensive intra/inter-LATA interconnecting for private-to-public interconnections. It will enable customer choice that ensures that broadband switching capacity is available and allocated on a dynamic "real-time" basis (class 5—metro switches).

So it is. We are now at a crossroads in time. We have a chance to build an exciting, growing, new, information-communications network for tomorrow, or we can achieve instead a complex, add-on, overestimated, underexpensed, limited voice communications network that might be difficult and expensive to maintain and support when used as the basis for future information services. We are going to do something anyway. We might as well do it right. In any event, the choice is ours. (See Fig. 8-6.)

LAYERED NETWORKS' ...
LAYERED SERVICES ...

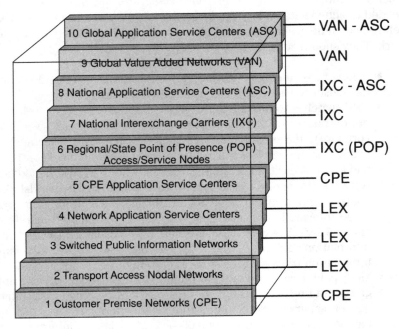

Layer	Service
10 Global Application Service Centers (ASC)	VAN - ASC
9 Global Value Added Networks (VAN)	VAN
8 National Application Service Centers (ASC)	IXC - ASC
7 National Interexchange Carriers (IXC)	IXC
6 Regional/State Point of Presence (POP) Access/Service Nodes	IXC (POP)
5 CPE Application Service Centers	CPE
4 Network Application Service Centers	LEX
3 Switched Public Information Networks	LEX
2 Transport Access Nodal Networks	LEX
1 Customer Premise Networks (CPE)	CPE

Fig. 8-6. Stepping along.

Visionnet

After all that has been said and done about new information network services, what does all this mean?

Assumptions

- New user requirements now need low-, medium-, and high-speed data exchange. Seventy percent are satisfied by narrowband, medium-speed, fast-connect, switched "any-to-any" systems, 20% by wideband, and the remaining 10% by broadband. Broadcast, providing entertainment-type services "one-to-many," can be delivered by cable, direct broadcast satellite, or UHF/VHF radio techniques.
- Serving the 18 major industries' "communities of interest" will be the best initial application of data, imaging, and video.
- Transmission bandwidth, computer power, and memory systems are becoming cheap. Software is and will continue to be expensive and complex.

- Public communications must be priced for growth. Otherwise, the trend for leased facilities for interconnecting to internal private networks will shift to include external private shared networks (VANs).
- Overlay will be the methodology for pre-1999. New integrated systems will be needed in 1999–2010.
- New revenues from new narrowband/wideband services can be used to help achieve the more ubiquitous new broadband infrastructure.
- Massive retraining for voice-grade data modem customers is required in order to achieve the new separate public data networks.
- Suppliers need three to five years of development time before extensive new broadband products are available. This will only occur when the LECs or IXCs or VANs step forward with a conceptual plan and begin working with suppliers to achieve an agreed-upon requirements specification.
- Planning today's and tomorrow's projects will require a shift in organizational structure to preprogram, preproject, and preproduct planning.
- As we consider when new services such as the class services, 800/900 database access, data handling, videoconference, and video gateways can be achieved, there are several techniques for producing new services for voice, data, and video. In considering where and how these services are obtained in terms of overlay, integrated, standalone, adjunct, signaling, call-processing interrupt, new generics, new systems, new software packages, new version releases, telephone-company programmed, and vendor-programmed possibilities, it is essential to ensure that funds are spent on a building-block approach that enables minimal impact on the network and maximization of funds available for next-generation systems. It is also important to recognize that the initial penetration will simply be the opening of a floodgate of increased usage. Then many new services will be layered on top of each other, requiring a new infrastructure to support the new movement.
- The network structure needs to be established so that on a phased basis, the narrowband, wideband, and broadband networks are properly overlaid to foster usage and demand. On the other hand, the actual permanent structure for movement of integrated voice, data, and video will be achieved by a new family of integrated switches, upon which layers of network access and transport services are provided.

Actions

The U.S. information transport providers need to first capitalize on existing plant for obtaining needed revenue, and then phase it out in favor of the more fully competitive broadband networks. To do so, the following actions and tactical strategies are both logical and implementable:

- Use ISDN as a key tool to shift from voice to data solutions by deploying ISDN in a large statewide area to make it ubiquitously available to all sections of the marketplace at a very economical cost.
- Deploy the narrowband data networks with their various options. Encourage the transfer from dial-up voice data to these new networks, deploying new transport services on an economical basis.
- Deploy data application service centers to provide the necessary extended and enhanced data transport, and database access services, that are required to be successful in the marketplace.
- Interconnect networks across LATAs to the world using interexchange data carriers at reasonable rates to encourage usage across states, regions, and nations.
- Overlay new data networks on an area-by-area basis in an orderly program over the 1990s.
- Ensure that the complete community of interest is addressable by the data service offering.
- Train marketing and technical people to fully understand data solutions.
- Deploy new data networks to impact every aspect of society (showcase future offerings as well).
- Participate in both the public and private arena to ensure that the CPE side of the network is addressed and resolved.
- Participate in the data standards arena, public and private, through IEEE, ISDN, EDI, and DOD.
- Promote data user groups to obtain support, understanding, and new requirements.
- Foster alliances among communication providers and computer and terminal equipment manufacturers.
- Obtain state regulatory relief and understanding to encourage implementation.
- Involve provider communication firms in preproduct planning to ensure that quality and the desired features are obtainable from supplier systems.
- Market the public data network dramatically as a dynamic new network that is available with a whole host of new data-handling services to encourage new-user participation. Teach them how to use it via formidable advertising.

Then, "Begin counting substantial, growing profits after the second year of operation."

The plan

Once these strategies are understood, the providers need to ensure the dynamic shift takes place by:

- Selecting major cities to deploy ISDN switching nodes to access major large-business customers, small-business customers, and residential customers over the late 1990s. Therefore, ISDN will be available to potential key users. For example, of the areawide customers in each of our major cities, approximately 25% of those would be ISDN data users. Extend the offering over a 60- to 100-mile radius of each major city.
- Fostering data mail service, alarm-monitoring services, energy management services, meter-reading services, health care information exchange services, remote teller radiology, police 911 caller ID data, electronic ticket purchases, and videotex-type offerings from a new host of suppliers over the data network.
- Ensuring that computer workstations and terminals are ISDN-compatible.
- Providing NT1, NT2, and TA network-terminating equipment at reasonable cost to foster usage.
- Establishing provider price tariffs with the state regulators to ensure usage at reasonable prices, such as $30 per month for the data network.
- Establishing market units and technical groups to work with users and demonstrate data solutions using the public data network.
- Providing, where ISDN is not yet deployed, a second pair of wires to customers to directly access standalone data switches that switch data users to the ISDN public data network offerings without a distance-penalty charge.
- Establishing C&C alliances between communications and large computer firms to enable large mainframes to provide direct access to the switched public data network, and to achieve application service centers that use advanced computers to meet various "data-handling" needs.
- Establishing program management groups to oversee providers' infrastructure deployment to enhance the movement and access of data.
- Establishing national and global networks with Data IXCs and VANs to interconnect the area to the rest of the world.
- Providing access to national and global application service centers.
- Providing long-term return-on-investment incentives to encourage fresh capital for a new "data-handling" infrastructure.

Conclusion

Let's assess the public data/information networks from the perspective of what needs to be done to achieve the key building blocks of the new U.S. information networks' infrastructure.

Pathways to the future information marketplace

Path one   Maximize our existing narrowband network voice offerings by using new ISDN (access), SS7 (signaling), and IDN (integrated digital network) technologies to upgrade the current infrastructure and provide expanded offerings. These offerings could include: voice mail; calling-party ID; class services; selected call transfer, blockage, message, routing, and priority override; intelligent network 800; 900 database access services; special audiotex services; fast provisioning; and number change over the expanded public voice network.

Path two   Establish a new family of addressable narrowband data services over the existing copper facilities, using ISDN data interfaces to enable both the circuit- and packet-switched movement of data from terminal to computer, computer to computer, and terminal to terminal over the new public data network.

Path three   Provide an exciting group of addressable wideband services ranging from wide area networks that provide a variable number of 64-Kbps channels up to T1/T3 rates of 1.5/45 Mbps for data, imaging, and graphic display of information. Applications could include: X-rays for the medical industry, stock forecasting for the securities industry, CAD/CAM for the manufacturing groups, and WANs for internetworking LANs over the public wideband network.

Path four   Foster the movement of information between private networks and public networks by providing new high-speed data transport services such as dynamic bandwidth, bandwidth on demand, survivable transport, improved error rates, access to multiple national carriers, and international value-added networks.

This will be achieved by expanding our offerings to provide these services over new fiber-based public broadband access networks. These will use self-healing, survivable fiber deployment topologies with new switching centers closer to the customer.

Path five   Achieve new revolutionary broadband services such as high-resolution video services that enable visual imaging, desktop conferencing, high-quality graphic display, computer-to-computer data exchange, video displays for education, media events, image archives, and entertainment video services. These would be offered over a revolutionary new fiber-based voice, data, image, and video-switched network with automated support capabilities, called the public broadband network. It would also provide access interfaces to broadcast cable, wireless cellular, and PCS networks.

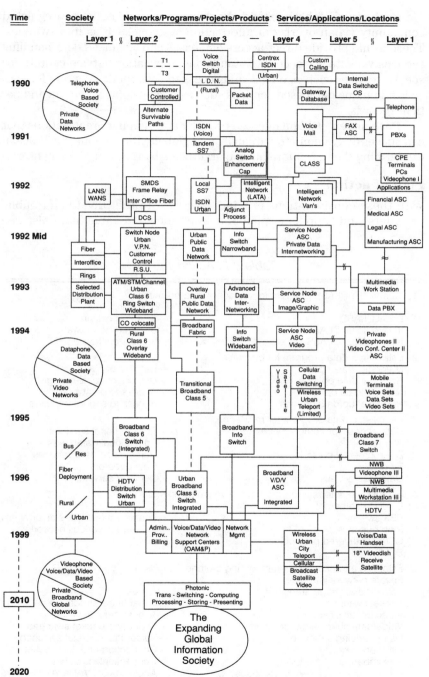

Fig. 8-7. The information game, the game of games.

Path six   Establish extended/enhanced services in the nonregulated community via an overlay of information systems such as info switches. These would provide direct-access databases and advanced data-handling services via gateway access to standalone application service centers for special services such as e-mail, and access to CPE service centers for specialized/unique databased services such as poison control centers and patient records.

Path seven   Obtain talented provider and supplier organizations that can successfully compete nationally and globally throughout the 1990s in establishing the new information marketplace by the turn of the century.

Plan of action

With these considerations in mind, let's turn to the private and public providers' and suppliers' "plan of action" shown in Table 8-2. (Also see Fig. 8-7.)

Table 8-2. Action plan

1. Establish a CLASS program for residence and business.
   —Number display
   Local CLASS
   Intraregional CLASS   —Selected call ring
   Interregional CLASS   —Selected call forward
                                           —Voice identification

2. Establish a data program for business.
   Major city data options   —Retail to banks
   Regional data options   —Doctors-hospital-druggist
   National data options   —Access to databases
   Rural dial-up/data options   —Access to supercomputers

3. Establish a data program for residence.
   Data access for home services   —Entertainment services
   Alarms   —Educational databases
   Environmental controls   —Home office data services
   TV display   —Data directory
   Database access   —Demand services

4. Establish an interface program for private networks.
   Extended ethernet
   PBX switched   —Enable various types of terminals
   Ring buses   to address various networks
   OSI/SNA layered protocol
   SONET

5. Establish a transport program for ring switches (CLASS 6)
   Private network interface
   Voice, data, image, transport   —Access CPE systems
   Packet switched   —Variable bandwidth on demand
   Video circuit switched   —Access service nodes
   Voice/data channel switched   —Wideband packet switched
   Usage pricing structure   —Voice/data channel switched
   Internetworking   —Connection-oriented packet
   —Numbering   —Billing   —Connectionless packet
   —Routing   —Access codes   —Access IXCs, VANs, ATPs, CAPs

6. Establish a development program for the service communication centers.

Voice/data/image switched packets	—Centrex features
Variable bandwidth on demand	—Data packet features
Distributed nodal point service complex	—CLASS features
Access to features/service complex	—LAN interface
Private network access	—Service node interface
Gateway to databases	—Gateway features

7. Establish a deployment program for switched fiber facilities.
   Rural/suburban/metro switches
   Ring access nodals
   Private network access

8. Establish national access service.

9. Establish international access service.

10. Establish usage pricing program for transport.
    Price for growth
    The poor have a nominal usage fee
    Charities obtain usage at protected rate
    Data options are priced for growth
    Variable bandwidth on demand
    Priced on service, not bandwidth

11. Selected open architecture access points.

12. Provide information services at application communication centers.

13. Formulate P&P partnerships among CPE and database firms for transfer and processing programs for "C&C," such as: IBM, Wang, DEC, Apple, AT&T, Sprint, MCI, Siemens, NEC, Fujitsu, RBOCs . . .

14. Establish PCS/PCN specialized network offerings for urban teleports (one way, two way).

15. Establish data/voice cellular network as inexpensive low-speed data offering to encourage mobile growth—wireless PBX, wireless centrex.

16. Automate maintenance and administration on public data network switches. Establish multiprovider/supplier programs to showcase voice/data/video maintenance and administration centers.

17. Establish a program for advanced intelligent network voice offerings.

18. Develop packaged features and services for global VANs' application service centers.

19. Automate internal business operations, using multimedia work centers to facilitate layers of software packages.

20. Establish a public data network directory.

21. Participate in Bellcore data and video network standard service planning.

22. Participate in the IEEE, T1, and CCITT standard meetings as well as ISDN user forums, etc. . . . .

23. Formulate partnerships among C&C firms for:
    Ring switches
    Operator support centers
    Info switches
    Metro switches
    Customer-premises information switches (I-PBXs/IBXs)

24. Adapt planning process across providers, suppliers, and users. Coordinate conceptual market and network systems planning for national uniform offerings.

25. Establish Internal MarkTec Planning Organizational Structures.

As the information network providers, working with their equipment suppliers, look toward the next 20 years, we could say that they are pursuing a five-step program as they implement their plan of action and venture down the various paths leading to the information marketplace:

- Step 1—Public switched voice network services.
- Step 2—Private overlay networking services.
- Step 3—Public switched narrowband data network services.
- Step 4—Public switched wideband network services.
- Step 5—Public switched broadband network services.

(See Table 8-3.)

Table 8-3. A five-step program

Step one—Public switched voice network services (1984–)

• Telephone	—Voice/audio service
• Dialtone	~2nd voice line
• Custom calling	~High quality voice
• Centrex/Centron	~Stereo 7 KHz audio
• Application service centers	~(D) channel signaling
—Fax (dial up)	—SS7 signaling—CCITT signaling
—Voice mail	system seven
• Digital upgrade	~CLASS
—IDN—Integrated Digital Network	• OAM&P upgrade
~Rural clusters	—Distributed administration
~Urban overlay	—Distributed operation
~Maintenance/administration	—Distributed maintenance
~Digital interoffice	—Automated service orders
~1 error in 10^7 bit/sec	—Rapid provisioning
—ISDN—Integrated Services Digital Network	—Rapid testing

Step two—Private overlay networking services (1991–)

• LAN to LAN interconnection	• Point to multipoint
—Frame relay	—Connectionless
—FDDI-I	—Connection-oriented
—FDDI-II	• T1–T3 hubbing
—SMDS—Wideband	• Video conference centers (private)
—SMDS—Broadband	
• Point to point	
—SONET transport	

Step three—Public switched data narrowband network services (1996–)

Network:

• Narrowband—ISDN	—Guaranteed error rate throughput
—Existing copper	—Data rate interfacing
—IDN-ISDN digital overlay	—Multirate interfacing
~Rural	—Blockage/reroute
~Urban	—Message sequencing
~Suburban	• Public data networks (Ckt, Pkt)

- Circuit switching (64K–128K b/s)
- Packet switching (9.6K–64K–128K b/s)
- Packet assembly/disassembly
- Ubiquitous
- Information transport services
  —Protocol conversion
  —Code conversion
- Alternate routing/attempt limits

—Addressing
—Routing
—Traffic loading
—Terminal interfaces
—Network interfaces
—Provisioning
—Test/support

Services:
- Data dialtone
- Dataphone
- Imagephone I
- Videophone I
- Data addressing
- E-mail
- Broadcast
- Polling
- Data digital fax group 4
- Delayed delivery/network retry
- Security
  —Encryption
  —Audit trails
  —Terminal access
  —Terminal ID
- D channel services
  —Alarm monitoring
  —Energy management
  —Meter reading
  —Home incarceration
  —Weather sensing
  —CPE to network signaling
  —ESP to CPE signaling
  —Terminal ID

- Integrity
- Privacy
- Multiple database access
- Closed user group
- Third-party software
- Record buildup via polling
- Classes of service
- Access networks
  —IXC—Data networks
  —ESPs
  —ISPs
  —Global data VANs
- Customer controlled VPN
  —Dynamically
  —Per call
  —Per day
- Application service center services—Access
  —Data storage/retrieval
  —Data manipulation
  —Data presentation
- Class of service
- Pricing
- Billing

Step four—Public switched wideband network (1997–)

Network:
- Access nodes
- Fiber deployment (selected)
- Customer variable bandwidth control
  —Setup
  —Dynamical
- IXC network interfaces
- P&P internetworking
- ATM/STM switching
- p X 64K bps—45M bps

- Fiber switched networking
  —Class 5
  —Class 6
  ~Sub 6
- Wideband OAM&P
- Protocol access interface
  —Ethernet
  —Token ring/bus
  —Frame relay
  —FDDI/SMDS

Services:
- Computer to computer
- Virtual private networking
- Closed user groups
- Variable "channel" bandwidth
- Wideband data address
- Dynamic bandwidth pricing

- Video conference centers
- Videophone II
- Imagephone II
- Multimedia/multirate workstation
- Security & survivability

Table 8-3. Continued.

Step five—Public switched broadband network (1999)

Network:
- Fiber-based—New topology
  - —Interoffice
  - —Selected distribution
  - —General distribution
- Narrowband-wideband-broadband inclusive
- SONET/B-ISDN interfaces

- User Network Interoffice (UNI)
  - —155 Mbps
  - —600 Mbps
    - ~ATM/STM
    - ~(N × OC-1) switching

- HDTV broadcast distribution
- Broadband IXC & VAN internetworking

Services:
- Video dial
- Video address
- Videophone III
- Workstation (B)
- Video conference center (public)
- HDTV
- Computer networking
  - —CPU—CPU
  - —CPU—workstation
  - —Application service center
- Video/image
  - —Video files
  - —Education
  - —Sports events

- —Entertainment
- —Search & browse
- —Image storage
- —Image manipulation
- —Image presentation
- —Broadband info switches
- —Closed user group
- —Global VAN gateways
- —Video storage access
- —Video mail
- —Integrated services
  - ~Narrowband
  - ~Wideband
  - ~Broadband

A

Future global information telecommunications

"I fear there will be no future for those who do not change . . . When there are no new ideas, things can remain the same, but strangers are coming with different ways . . . Growth comes from change . . . A people grows or it dies . . . There is but one thing we know . . . and that is that nothing forever remains the same."

Louis L'Amour

Differing views, differing perspectives

Different views, different perspectives, and different goals have led different players to different networks with different services, as they provide different solutions to address different needs. Over time, these differences and somewhat difficult choices have become more complex and more crucial, as the forces for the merging and integration of the computer and communication (C&C) industries intensify and become more massive and explosive.

To best understand and appreciate what is happening, it's time to step back from our active involvement in playing the information game and take a broad view. Let's encompass the full scope and range from not only today's perspective, but also tomorrow's. Let's review what has happened to bring us to this juncture in time. As we assess how we came to this particular game, we might want to reassess the game itself, redefine it, or not continue to play it at all.

Some might say, "What's the difference? All roads lead to Rome." Perhaps they did for the weary traveler some 2000 years ago, when one ruler could extend his or her influence over many diverse groups. But today, this is a different story. Today, there are many paths, and they all do not lead to Rome, or some Mecca or Shangri-la depicting success, fortune, peace, and comfort. Today, the issues have become multifaceted. In today's competitive arena, no single leader, such as AT&T in communications or IBM in computers, is alone able to set direction. Today, the human and C&C elements have grown quite complex. One might say that they "grew and grew until gruesome." We have entered a new arena, where C&C considerations now require interfacing, interlacing, interconnecting, integrating, internetworking, interprocessing, and interservicing.

For many, it has been difficult to put their arms around the key strategies needed for successful play. Some have simply moved from game to game, as though at the Mad Hatter's tea party, leaving things in a worse state than before. So what has happened to the national information networks and services that many believe will have a monumental effect on our society, bringing about great social, economic, and cultural change? Indeed, what has happened to the missing information infrastructure? What is happening today? What could, would, should happen to achieve it in the not-too-distant future? What are the prospective users' needs? What are their views?

A user's PCS view

For the past year, as a potential new user of personal communications services (PCS), I have conducted a personal experiment on wireless communications. It is well known that telephone executives use cellular phones more extensively than the average businessman or woman because of the vast number of people they manage and the fact that they are telephone people. When a conference breaks for a moment or two, most participants venture out for a breath of fresh air and relief, while the telephone people seem to run to the pay phones. Thus, it is quite natural and understandable that one telephone executive's business card noted an extensive list of phone numbers to enable him to be reached anywhere. This occupational "love affair" with the telephone, desiring continuous communication, then led to the extensive push that is currently prevailing in the telephone industry for delivering a personal phone.

As the FCC has wrestled with spectrum availability, the technologists have wrestled with CDMA versus TMDA or extended TMDA feasibility, the marketers have wrestled with unit size and leased versus ownership prices, and standards groups have wrestled with interconnect interfaces for multiple providers. Meanwhile, I have wrestled with concerns of the usability of PCN and the possibilities of "tulip mania," where the Dutch created a rush for a particular bulb that was not accepted externally, thereby causing numerous bankruptcies.

When one considers the tremendous transport capability of the fiber and its tremendous costs compared to alternative wireless and cable technologies, it is quite apparent that we do not have the capital to do everything. Now is the time to put on our user hats and play with these diverse offerings to identify their real limits and boundaries, to better assess their scope and potential, and to better evaluate conflicting strategies and alternatives.

With this in mind, I set out to determine if indeed the phone in the purse or pocket was all that we in the telephone industry believed it to be. Are we indeed simply creating a market in our mind, while the majority of prospective customers will not be as equally keen and willing to part with their hard-earned money. It is a small, modest attempt to determine what additional features, if initially provided, would help the personal phone (PCN/PCS) to be accepted.

So, in the past year or so, I used both the "brick" and "slimline" versions of the cellular phone to determine the pros and cons of actual wireless usage, at least from the perspective of a hopefully somewhat unbiased user. To be fair, I am sure many will agree or disagree with these observations.

The initial user test

The initial user test was to be exactly that, a randomly generated, real-life experience in which the user simply carried the device everywhere. It was not confined to the car. It was to be continuously with the customer. I put it in a special bag, which I carried wherever I could as conveniently as possible. During the course of everyday events, I came to the following observations and conclusions.

There are indeed times that even the most resistive household members would like to have a personal phone available for contingencies or emergencies. At these moments, I would have paid considerably to have mobile-phone capability:

- A night ride through the country alone on bad roads during the winter.
- The need to call an associate who does not arrive at a predetermined destination, where no local pay phones are readily available.
- Emergency conditions when a child is being raced to a doctor's office only to find that the doctor is not there, but at the hospital.
- Times of personal illness, where one could call in a change of plans while actually proceeding to change them without delay.
- The ability to establish a dynamically changeable meeting place and time depending on traffic (for example, asking a person to meet you with a document when you are actually within the area).
- The ability to dictate a letter while driving to work or waiting in traffic, thereby already having a first pass available upon arrival.
- The ability to have a selected call forwarded to you when you leave the office en route to another location.

- The ability to easily call work or home when in a distant location or city. The more foreign and remote, the greater the appreciation of being able to use a familiar device to quickly and easily reach the home base for information or messages or to convey status.
- The need to receive calls at a remote location, especially to resolve emergencies and crises or to provide status to those behind.
- The need to receive calls at remote locations that change dynamically (for example, as one's car travels along a predetermined path across the country).

These applicable needs—and others that I am sure other participants could identify—do indeed make a case for having a personal phone. However, it is important to differentiate the reality of actual operation from these general needs and specific requirements. This helps establish the personal worth or value that the user will place on the actual use of the service. In this way, we might help determine potential opportunity for differently priced offerings.

In actual usage, the following was noted:

- The weight and size did directly affect where and when the device would be carried, but need many times offset negatives, depending on the seriousness of the situation and the desire to receive a pending message or communicate a situation or status.
- There is a need to easily communicate with the instrument to direct the call to a desired destination and receive calls, expected or unexpected.
- There is the need for reliably receiving and transacting calls with appropriate power and transmission reception.
- There is the need for the power on/off button to be extended and made larger so it can easily be found in a darkened situation.
- There was inability of initial units to use outside batteries, such as a car cigarette lighter. This caused extensive reliance on internal batteries and required changeable batteries and low battery features.
- In most cases, the low-battery audio alarm does not work. In ten cases out of ten, the alarm beep was first heard as the unit immediately shut down.
- Never knowing how low the batteries really were, constant recharging of batteries was required.
- Incorrect charging caused extremely short life of the batteries, as recharge did not take unless batteries were fully depleted. This led to a myriad of problems and situations, causing the user to question the reliability of the phone.
- The inability to call properly during certain weather conditions or dead spots caused conversation quality to deteriorate or end within a city or remote area.

- The ability to have selected calls forwarded could only be achieved manually by a secretary. When the secretary was not present, calls were not forwarded. Hence, reliability was in question.
- Inside parking or building structures inhibited calls, causing numerous dead periods. This usually happened, as Murphy's Law would suggest, at the exact time when an important call was attempting to reach the roaming party.
- Even upon establishing one's location in a distant city immediately upon arrival, as the car passed through the major city and/or as the user stopped for a break, the cellular roaming locator was not updated fast enough in "real time" to locate the party. Hence, incoming calls could not reach the party.
- A remote-city stay required many hours to update remote-call transfer rerouting capabilities. In most cases, this period was more than the length of a short day visit.

The movement of data on these wireless systems to and from distant cities would be feasible, but reliability and privacy is highly questionable. However, broadcast data/video services are, of course, quite acceptable.

The cost of remote calls is prohibitive. When all is said and done, remote-call costs are expensive because all the fingers in the pie want a piece of the action. The worst offender is the cellular provider on the airplane, whose charges are in excess of $6.00 or so per minute.

Conclusions

We should remember that cellular was tariffed and priced to the FCC as a service for a limited group of users. This was due to spectrum considerations. It was to be for the executives, the mobile salesperson, the small businessperson who operates out of the truck. It was not for the masses and not priced to encourage massive use anywhere, anytime.

On the other hand, the more universal needs are indeed there. However, the personal phone does need to have the ability to be disconnected, as there are periods when we do not want or need to be reached anywhere, anytime. Thus, a "Dick Tracy" personal phone would be a nice luxury. In some instances we could even say it is essential, but price and availability will be the determining factors. Once we begin using something, we shift our mode of operation to use it more, and we become more dependent on it. Then it requires more and more reliability, portability, and quality, which some call availability.

For the masses to use it, the spectrum size and weight problems need to be resolved. In many instances, the handset (even mounted in the car) is dangerous, as in dialing numbers while moving. To truly have a safe offering, we need to be able to talk to the terminal, whose specialized personalized voice recognition technology enables the singular user to activate connections verbally. This, together with a much better battery status/charge feature, would greatly help mass usage.

Finally, the roaming and call-transfer features need to be carefully addressed to resolve the operational capabilities and make it versatile and functional. Only when these items are addressed in a cost-effective manner will we truly lead the customer from the 25-cent pay-phone alternative. This might be impossible when we consider the fully operational costs involved in deploying a more ubiquitous service that totally achieves the anywhere, anytime objective.

A user's ISDN needs

As noted throughout the preceding analysis and in my earlier work (*Future Telecommunications*), there are numerous (32 or so) tasks for achieving a narrowband public ISDN-based telecommunications network for delivering moderate-speed-data and video-image communications. The tasks note the need for ubiquitous deployment at least throughout the major cities and rural county seats so that small businesses have the ability to network information and access large businesses and governmental databases. Similarly, 10% to 20% of the residences desired "data networking" capabilities. These enable doctors to review X-rays from their homes, managers to access company files, and marketing and design personnel to remotely perform their work via telecommuting.

To achieve success in ISDN, it is therefore necessary to successfully implement each of these tasks. Depending on the degree and extent of availability, we need to implement them for narrowband offerings, and broadband ISDN networks need to follow a similar methodology for their success with deployment.

From resolving ergonomic interface needs to providing educational advertising on how to use the services, several issues need to be resolved: pricing, addressability, CPE, robustness, ONA standards, network management, and deployment availability (locally, regionally, nationally, and globally). These, along with customer understanding, appreciation, and acceptance, will foster ubiquitous use of ISDN technical possibilities in the forthcoming information marketplace.

Pricing

To encourage universal acceptance and use of narrowband ISDN, we need to immediately address ISDN pricing for a public data network offering. Narrowband ISDN delivers three networks:

- 64,000/128,000 bps data circuit switched transport.
- 64,000/128,000 bps packet-switched data transport.
- 9600/16,000 bps packet switching data and signaling transport.

These networks need to be deployed ubiquitously throughout an area and priced so communities of interest will use them to exchange the full range of information transport. Hence, ISDN should not be sold as an interface. It should be sold as a series of networks that support higher-level services priced to attract users to the new data-handling networks and away from dial-up voice-grade data modems on the existing flat-rate voice-service network. Some believe these prices need to be tiered, depending on use, so that initial offerings are priced as low as $16 per month for certain amounts of use and local area-wide access.

We also need to deploy switched wideband and broadband offerings so that videophone, for example, and high-definition videoconferencing and television can be economically accessible to all. This will shift our pricing to being based on bandwidth and not simply relating it to the equivalent number of voice conversations. In this manner, videophone could be available ubiquitously for $30–60 per month. As we have found in Germany, we must be careful in separating these narrowband and broadband services to ensure that each does not inhibit the other from flourishing. There is a need for considerable work in this regard to ensure that an appropriate new pricing structure is effectively established to ensure growth.

Addressability

Only when there is a similar available dataphone directory equivalent to the telephone directory will full data movement be achieved. The RBOCs, independent carriers, and private carriers must work together to obtain interconnectability and a universally available address directory for our customers. This is key to the success of public data networks.

CPE

As today's customer-premises equipment embraces ISDN chips to replace low-speed modems, there is a need for these systems to change out their transport protocols to new protocols based upon these higher-speed, less-error-rate, digital-data ISDN network capabilities. As rate adapters and inverse multiplexers enable asynchronously disparate terminals to access these higher-speed synchronous ISDN networks, we need to develop more sophisticated terminals that use functional/stimulus "D channel" network interfaces to enable more sophisticated network and CPE operations.

Robustness

Computers have entered every aspect of the marketplace, and we need to encourage computers to interconnect and exchange information. Thus, we need to prepare a network that can successfully handle more and more traffic. The worst scenario would be to encourage the hospitals, doctors, police, government, and other society-supporting agencies to shift from private networks to a public data network that cannot handle these increased traffic loads. When throughput loads for increased traffic reach unattainable network operational objectives, they go down and cause disruptions at crucial times. This occurred in France and caused three days of data network outage. This would be similar to encouraging people in Los Angeles to use their cars so they can sit in traffic jams on freeways that cannot handle the growth in traffic. It would be better to have automobile transport and trains as well. Similarly, the shift from narrowband to wideband to broadband must be timed appropriately to facilitate growth.

ONA standards

Past endeavors in ONA have reached stumbling blocks as well-meaning parties attempted to open up the switching systems so that multiple providers can provide multiple services without being inhibited by cumbersome procedures, standards, and delays. To obtain a competitive marketplace, it is indeed essential for numerous, diverse services to successfully grow and flourish. Thus, an information networking architecture is needed to provide a desirable standard interface for exchanging services so that one system can offer customers to another system for enhanced information networking and layered service offerings using the layered networks' layered services model defined in *Global Telecommunications*. This model enables functional separation at universal interfaces. Each layer can enhance the next for both private and public offerings, enabling not only LECs and IXCs, but also VANs, VARs, CAPs, ATPs, and SPSs to participate effectively without danger of bringing the network down. This model needs to be challenged and used to its fullest. Numerous firms throughout the world are now exploring not only its local ONA interface possibilities, but also its global networking aspects.

Network management

When all is said and done, it is essential to be able to manage the movement of information. In reviewing the layered networks' layered services model, it is quite apparent that CPE systems require the model's network management capabilities. Local exchange carriers (LECs), interexchange carriers (IXCs), and value added networks (VANs) need the capabilities as well. So as we enter a more global community, the global networking aspects must be integrated into our network management hierarchies. We must construct appropriate network management mechanisms for today's voice traffic and for data and video as well. Thus we must resolve who the "keepers of the network" are. Who will ensure that the layered networks' layered services model can be successfully implemented? Is this a task for Bellcore? Is this a national standards challenge? How can we ensure that multiple service providers all function together appropriately? This is indeed the challenge of democracy, where everyone can do their thing without affecting the freedoms of others.

Availability

Availability will not be an issue once these questions have been resolved. RBOCs are very good at deployment once the risks are reduced and the customer's needs are understood. However, there is a need to shift emphasis from short-term, fast-revenue nearsightedness to full-service, full-deployment, long-term visions. Providers need to address not only the narrowband, full-deployment strategies using current upgraded plant, but also the more expensive fiber-based deployment strategies required for high-resolution, high-quality video.

While addressing these particular key issues, market units need to concentrate on the customer. They need to educate the customers on the technical possibilities that will be available to them. As noted in the many implementation tasks, this will require extensive advertising to clearly demonstrate what the technologies offer and how the user can employ them to obtain a better quality of life. We, the providers, suppliers, and users, need to work together—all of us together! The information service providers (ISPs) and enhanced service providers (ESPs) cannot offer their services if these public networks are not available. We need to ensure that private networks flourish as they use shared public offerings to achieve their individual tasks. Using front-end planning processes, we need to work together to determine where we are going and how we will get there. For these reasons I have written a series of books for the industry addressing this forthcoming information marketplace and showing how we can work together and what needs to be done.

A different perspective

A final look at the telephone, dataphone, videophone, and personal phone

Several analysts were asked, "What should we be doing differently from what we are doing?" The answer was that it is time to reevaluate what we are trying to accomplish today for tomorrow's information society. Do we want to achieve a new data-handling network, similar to the voice telephone network, but specifically for data, so a dataphone can talk to any other dataphone, anywhere, anyplace, anytime? What are we doing to deploy videophone in its narrowband, wideband, and broadband forms?

One network watcher noted that the ISDN data networking game took a serious turn in 1992, as the ISDN Executive Council for the Corporation for Open Systems International (COS) launched National ISDN-1 (NI-1) to resolve ISDN connectivity between different vendors' central office switches. This alleviates the proprietary, preubiquitous ISDN implementation problems, where CPE terminal equipment, working well with one vendor's network switch, required a different programmable ROM to work with another. Prior to 1991, the number of access lines equipped to deliver ISDN was only

200,000 or so. By the end of 1992, Bellcore estimated that ISDN was available to approximately 62 million lines. NI-1 addressed many of the analog needs for narrowband BRI, while wideband PRI, at the local level, is addressed in NI-2 and NI-3 software releases. The following is a list of software release dates: NI-1 was available by AT&T 5ESS Software Services 5E8, May 1992; Fujitsu Network Switching of American Fetex-150 Release G3, February 1992; Northern Telecom DMS-100, BCS 34, September 1992; Siemens Stromberg-Carlson EWSD, Rel 10, November 1992; and Ericsson AXE, Rel 6, March 1994. While local exchange carriers (LECs) have pushed the deployment of BRI, interexchange carriers (IXCs) have opted for PRI. LEC-BRI is essential for the CPE, and centrex offerings. PBX vendors use BRI on the line side and PRI on the trunk side as they connect directly to IXCs.

Trip 92, coordinated by the Committee for Open Standards and Bellcore in November of 1992, achieved the "golden splice," marking the entrance to the information age. Coast-to-coast ISDN communications were achieved, connecting 20 cities and providing numerous ISDN voice, data, image, text, and video demonstrations for education and medical imagery applications. The "golden splice" is similar to the beginning of the Industrial Age. When the coast-to-coast railroad track was finally completed, builders drove a "golden spike" connecting tracks from both the east and west coasts. So this "golden splice" similarly marks the linking of ISDN's coast to coast, and the entrance into a new age—the information age.

The North American (N.A.) Users Forum, in Gaithersburg, Maryland, participated in identifying some of ISDN's most valuable applications as:

- Videophone and videoconferencing.
- Telecommuting.
- ISDN telephone/ISDN workstation integration.
- Multipoint screen sharing.
- Customer service call handling.
- ISDN access in geographically remote locations.
- Image communications.
- Automated number identification.
- Multidocument image storage and retrieval.
- Multimedia services.
- Multiple ISDN telephones on a single BRI loop.
- ISDN interface to cellular, mobile radio, and satellite systems.

Several issues were noted, such as tariffing and provisioning of ISDN, where ISDN's service negotiation procedure involves hundreds of subscription parameters, such as those for billing, location, and function. To reduce this, a special grouping of ISDN subscription parameters now enables a number of ISDN parameter groups and settings to be deployed.

It should also be noted that CPE stimulus/functional protocols for working with the network to accomplish a particular service is also a concern, as NI-1, 2, and 3 solutions are delivered. It is also important to ensure that compatibility with existing solutions is also resolved with national ISDN solutions. NI-1 relies extensively on network intelligence embodied in service protocols, which are also known as stimulus protocols. For example, to achieve call forwarding in NI-1, a user would dial an access code or hit a feature indicator. The CPE would essentially sit on the sidelines as a passive entity. However, more intelligent functional protocols would let the CPE interact to a greater extent with the network. The CPE would then understand and use protocols different from stimulus protocols that enable it to functionally intermix with the network to achieve the designated task. These are known as functional protocols.

SS7 signaling fully implemented with ISDN will be required to achieve the whole host of new ISDN supplementary features that the "D" channel can deliver with intelligent CPE. Calling-line identification, call hold, selective call forwarding, and voice to text features will require considerable support and interoperability as defined by Bellcore's "General Guidelines for ISDN Terminal Equipment on Basic Access Interfaces" document. This document needs to be further detailed throughout the 1990s to resolve ambiguous areas, expand "D" channel features and services and address terminal compatibility, data networking, CPE testing, power requirements, and maintenance of ISDN CPE. As NI-1 became available in 1992, NI-2 in 1993, and NI-3 by mid-decade, they continued to address consistency, connectivity, and cooperation issues and concerns, for ISDN is as yet "not too late for the data party."

One observer noted that the most popular video application relying on these new switched digital services is national and international videophone and videoconferencing. Videoconferencing users typically rely on multiple 56 Kbps or 64-Kbps channel services, known as N × 56/64, or on 384 Kbps digital services. Videophone, at 2B or 128 Kbps, has tremendous possibilities.

Switched data services are also increasingly being used for:

- Replacement of private line Dataphone Digital Service up to 56 Kbps with switched digital connections.
- Extension of high-speed data links to small, remote locations where leased lines have been uneconomical.
- The internetworking of geographically separate LAN traffic—on a demand basis—through the use of intelligent bridge/router/hub/node controllers.
- The addition of capacity to existing leased circuits on demand during traffic peaks or when unanticipated demand grows beyond leased-circuit capacity.
- For redundancy, backup, and disaster recovery when DDS and private leased lines fail.

While waiting for ISDN, various providers, such as AT&T, MCI, Sprint, Wiltel, and the RBOCs, are deploying numerous data handling "virtual" data networks and providing many private network services to encourage users to shift from direct leased lines to shared, usage-based facilities. These facilities offer the following: 56/64 Kbps, $n \times 56/64$ Kbps, and fractional T1—384K and 786 Kbps.

Several data transport services have developed as "rate adaptation" techniques. The services use V.110 or V.120 to enable different subrates to use ISDN BRI or 56K/64-Kbps switched facilities. Single application "inverse multiplexers" are just the tip of the iceberg in the world of public data network services, previously occupied by channel banks and T1 backbone networks. Next, "continuous bundling" of 64-Kbps channels to the 384K, 1.5, and 2.08 rates compared favorably with the alternate reverse multiplexing techniques, enabling dynamic selection of multiple BRI channels to fulfill needed bandwidth requests. "Continuous bundling" takes only 2.5 seconds, compared to reverse multiplexing's 30–45 seconds of setup. Working with the switched digital services application forum, Bandwidth On Demand Interoperability Groups (BONDING) has addressed inverse multiplexing interoperability issues. Numerous firms, such as Ascend Communications, Digital Access Group, Newbridge Networks, TynLink, and Axxess, telephone companies, and Prompters Communications, Inc., are engaged in deploying their product solutions to inverse multiplexing.

Also, ISDN's setup time, typically less than one second, compares favorably with traditional 10- to 20-second dial-up connections. This is such a significant reduction that it justifies using ISDN over alternative dedicated T1 facilities, depending on traffic volumes.

Thus, AT&T's Software Defined Data Network (SDDN) offers N × 56/64 Kbps and switched 384 Kbps, as part of their Switched Digital Services. MCI began offering switched data services in 1991 under its VNET virtual services umbrella. MCI's virtual data services family includes switched T1, DS-3, 56/64-Kbps, frame relay, and continuous 384-Kbps service. Sprint's Virtual Private Network provides VPN56, Switch 64, and 1.536 Mbps. Both MCI's Digital Gateway Service and Sprint's Integrated T1 Access Partitioning lets users rely on a single access line for private and switched transport. Similarly, PRI services will enable direct access to "B" channels within the 1.544 bit stream, as they compete with T1 leased-line offerings. These data speeds have taken a quantum leap in magnitude from the 2.4K, 9.6-Kbps top-of-the-line V.32 modems.

PRI is appealing in that, unlike T1, individual B channels in a PRI can be dynamically reassigned to different carriers on a call-by-call basis. So it's a highly flexible alternative to T1 for access to a diversity of carrier services, such as switched 56 Kbps, WATS, and end-to-end ISDN B-channel data calling such as voice data, group 4 fax, or LAN interconnection.

One analyst observed that, besides these data transport operational characteristics, the spectrum of ISDN BRI equipment repeats a cornucopia of capabilities. These range from devices that contain highly sophisticated video cameras and compression algorithms, which use one or more switched ISDN B channels for carrying full-motion videoconferencing, to those confined to performing LAN gateways, such as remote Ethernet LAN bridges. These use the very fast call setup times of ISDN to dynamically establish a high-speed data telelink to forward data when requested and then just as rapidly drop the connection. (See Fig. A-1.)

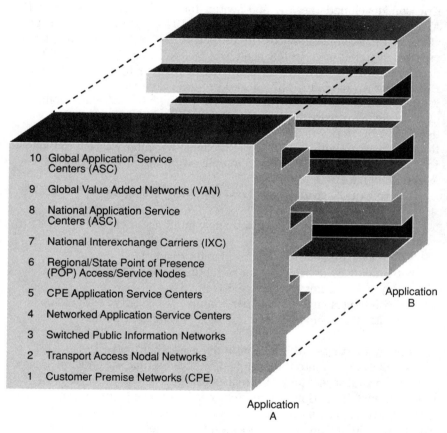

10 Global Application Service Centers (ASC)

9 Global Value Added Networks (VAN)

8 National Application Service Centers (ASC)

7 National Interexchange Carriers (IXC)

6 Regional/State Point of Presence (POP) Access/Service Nodes

5 CPE Application Service Centers

4 Networked Application Service Centers

3 Switched Public Information Networks

2 Transport Access Nodal Networks

1 Customer Premise Networks (CPE)

Application B

Application A

Fig. A-1. Who keeps the structure together?

New switching systems

The role of CO switching is changing dramatically, as noted in *Global Telecommunications* (Heldman, 1992, McGraw-Hill). That book reviewed the history of past switching systems and covered the progress from electromechanical to electronic wired logic to stored program control (STP).

In 1878, American District Telephone Company of Chicago began telephone operations, having the first private telephone system in New Haven, Connecticut, which was created with 21 subscribers and a manual switchboard. In 1889, A. B. Strowger developed the step-by-step system that was installed in 1892 in LaPorte, Indiana. In 1900, the Rochester rotary system was introduced from Stromberg-Carlson. In 1915, the Panel System was introduced and installed by AT&T. In 1916, Bell adopted the Strowger Switch. Late in 1924, Bell Labs redesigned the crossbar. In 1930, L. M. Ericsson developed the ME 500 switch for small rural areas. In 1941, AT&T's No. 4 crossbar was deployed, and in 1948, No. 5 crossbar was provided in Philadelphia suburbs. In 1955, the EMC rotary switch system/crossbar switches were widely used by AT&T No. 1 and No. 5. In 1960, the first electronic wired Logic Switch was developed by GTE, Automatic Electric for Autovon. In 1967, Stored Program Control switches were developed by Bell Labs. In 1976, Digital Switching No. 4 ESS and common channel signaling were developed. In 1977, Digital CLASS 5 switches were developed by Northern Telecom, followed by, in 1978, GTE with Digital Remote Switch Units. In 1991, Broadband Switching was developed by Fujitsu, and in the early 1990s, photonic switching prototypes were created by Bell Labs.

The role of CO switching is, however, finally changing:

- The line functions need to exhibit multimedia capability closer to the end user.
- Services are database managed and provided.
- Signaling creates intelligent communications.
- Support systems are service driven.
- Switching is on a per-call, per-channel, per-facility basis, with bandwidth defined in line units and billed in the central office.

But capitalizing on this discontinuity is another matter.

As ISDN moves into the switched multirate arena, as enticing as new switched digital access might be, real excitement can be generated with devices that match multiple network choices in a single platform. These new network access servers must seamlessly integrate and access both dedicated and switched digital networks, since the objective is still to match the application with the best possible network service. What is required is an integrated platform of connection technologies that provide access to dedicated and switched digital networks with consistent user control, interchangeable interfaces, and each under a single network management system. This is the concept of dedicated and switched digital access (DSDA, noted by industry watchers). Variable bandwidth can revolutionize network optimization because, for the first time, applications can truly be matched with the appropriate network services.

In addition, newer switched digital servers can provide access to both dedicated digital networks and switched digital services over the same local access facility. In this way, local access charges are minimized. Digital access nodes are particularly well-suited for combining dedicated and switched digital services via channel switching n-x-64-Kbps methods. Channel switching will be the preferred method of LAN-to-LAN connection. Variable, high-speed digital connections will meet the future requirements of LAN-to-LAN traffic. During peak transmission periods, incremental bandwidth can easily be added and removed as traffic patterns vary. When tariffs permit, similar flexibility can be applied to broadband ISDN networks. Intriguing possibilities emerge when inverse multiplexing is combined with T1, frame relay, and ISDN. Multiple switched 384-Kbps or switched 1.536-Mbps calls can be combined for efficient transport, and could present an alternative to switched multimegabit data service. Multiple T1 calls can be established to back up T3 backbone networks. Single access to private line, frame relay, and switched digital networks is also possible. The right combinations of technology solutions and lower network prices can now provide users with a continuing array of network alternatives. By combining the method of access to both dedicated and switched networks through standardized interfaces, the communications professional can match applications requirements with the best possible network alternative . . . so it goes.

We are moving rapidly from a traditional AT&T-based network to a global ISDN-based network with distributed switches closer to the customer, providing a whole host of new, exciting information services in the new global "information telecommunications" marketplace.

A final look notes that real change requires real commitment to address real issues to make real decisions and take the real risks to achieve real successes. Hence, different visions from different perspectives provide different networks and different services, as different providers, making different choices, pursue different paths, where fortunes and destinies are made or lost by these differences.

> *Who is to say*
> *What the future will hold,*
> *What path will be taken*
> *Where, when, by whom.*
> *The answer is . . .*
> *blowing in the wind.*
> *The answer is . . .*
> *blowing in the wind.*

Internet

A precursor to the public data network

Peter K. Heldman

Data networking history

Before looking forward to the future, let's first pause and look back at where we have been and determine how we got here. This will help put current and future events into proper perspective.

Over the 1950s and early 1960s, military frequency shift key start/stop data networks and commercial Western Union telegram facilities transported information at teletype/telex rates of 60 words per minute (wpm) and 100 wpm, using field data code of 5 bits per character or the ASCII code of 7 bits. After the antitrust suit in the mid 1960s between AT&T and IBM, AT&T mainly supplied dial-up voice-grade modems operating at 1200, 2400, and 4800 bits per second. These 100/200/300 series data sets operated either on expensive leased-line facilities or on the voice network as dial-up data transporters. During this period, computers extended their batch processing to front-end/back-end, foreground/background fixed-position/variable-position processing so that multiple programs could be remotely inputted and outputted to and from the traditional large mainframe computer. As more and more front-end compilations (compilers) work was achieved by more sophisticated and powerful standalone systems, there was a need for interconnecting the local community using the extended bus structure, enabling messages to be sent to various terminals via an internal address envelope. This system evolved through the IEEE 802 standards (see appendix B) to become the local area network (LAN), and wide area network (WAN) transport mechanism.

In the 1970s, local area networks blossomed as they were interconnected first by bridges and then routers to transport messages to different floors in the building or across the local campus. IBM introduced new transport protocols (VTAM, SNA, and SAA), DEC introduced DECNET, and Wang had Wangnet. Wide area networks were basically direct leased lines across town or to centralized nodal switches on customer premises or dial-up voice-grade transport. Rates increased to 9.6, 14.4, and 19.1 kilobits per second. During this period, Department of Defense networks were enhanced and expanded, and universities began internetworking research information over packetized networks such as ARPAnet, where information was packaged in envelopes containing header, text, and tail information. From these humble beginnings in the 1980s, ISDN (Integrated Services Digital Networks) standards were developed to enable internetworking of data at 56,000/64,000 bits per second. The computer-on-a-chip technology made standalone personal computers more and more powerful as databases became more and more sophisticated. Without the availability of an ubiquitously available parallel public data network, Internet was born. It consists of basically a service node attached to a university's or government research center's local area network, interconnected throughout the country by digital T1 trunks using a global address structure so that each terminal had its own address. Those not on the local LAN could obtain access via a dial-up voice-grade modem operating at 1200, 4800, and 9600 bits per second.

So, in reviewing the past, it's obvious that there was a need to interexchange information but what was missing was a network specifically tailored to do so, especially one in which anyone's terminal had a recognized address and the network deployed a protocol such as TCP/IP, enabling the transport and acknowledgment of receipt or requests for repeated transport to ensure message delivery. These mechanisms came straight from the earlier navy networks, SAC communication networks, and Air Force 465-L, 490-L, and Autodin data-handling message systems.

During the early 1990s, as the world caught on to the dial-up voice-grade 2400/9600 fax network, where facsimile systems scanned and transmitted the entire page, all white and black areas, the shift occurred to e-mail, where text (or graphic) information was delivered electronically between personal computers. Thus sprang up the electronic bulletin board, where those in a closed user group could leave information for others to access. From this came the open Internet, open bulletin boards, enabling anyone to post anything or send anything to anyone. Hence, the market need was established for exchanging data. Let's now assess how connectivity and interconnections are indeed being achieved and supported today and whether there is, perhaps, a better alternative. Let's begin with a detailed, specific review of the birth and progress of the Internet.

Internet's history

In order to understand what the Internet is today, it's necessary to understand first how it has evolved. In the late 1960s, the Department of Defense (DOD) sought to link its various computer centers. The ARPA (Advanced Research Projects Agency) was formed, and by the late 1970s a wide area network (WAN) called ARPAnet, as well as other radio and satellite networks, had been established. In 1982, the military decided to adopt Transmission Control Protocol/Internet Protocol (TCP/IP) as their networking protocols, and UNIX as one of their main computer operating systems. Soon ARPAnet was expanded to include many military sites, effectively doubling the number of terminals it supported.

In 1985, the National Science Foundation (NSF) funded the growth of Internet to 100 universities across the United States, linking researchers and supercomputers to improve the distribution and communication in the research and science community and to improve access to supercomputing resources. Eventually an NSFnet was established, connecting five supercomputing sites, and this was linked to the Internet.

With this tremendous growth of users, NSF realized that Internet would soon be exhausted, so in 1988 they chose IBM (for hardware and software), MCI (for telecommunications services), and MERIT (for management and operation) to implement a new NSF backbone network with three times the capacity of the former network. By late 1991, this too neared exhaustion, and NSF realized that it was becoming too costly for the government to continue supporting Internet alone. So IBM/MCI/MERIT formed ANS—Advanced Networks and Services, a nonprofit corporation to build yet another Internet backbone, with joint commercial and NSF funding, which is currently used today. ANSnet has 30 times the capacity of the NSFnet backbone it replaced. Thus began Internet's transition from a federally owned and controlled public resource to a privately owned and operated commercial network.

Internet today

Today, the Internet is composed of a group of 15,000+ store-and-forward packet networks, supported by TCP/IP protocol. These networks are linked together across the globe to form a single virtual network. The networks fall into three broad categories:

- ANSnet—uses DS3 facilities, which link major supercomputing sites to form the Internet backbone.
- Midlevel/regional networks—link universities and research facilities regionally.
- Individual networks.

The Internet uses a series of cascading routers to link these various networks, which communicate using the IP protocol for networking and the TCP protocol to ensure reliability and integrity of the payload. Each network segment is individually managed but allows the free passage of information throughout its site as data is transferred across the country. This might soon change because NSF plans to implement a policy of blocking all but academic and government traffic that attempts to pass across its backbone. The intent is to protect its facilities from the Internet's new growth in commerical use.

The Internet has evolved to many new uses since its inception as a research and development (R&D) network. Today it predominantly supports three main applications: e-mail, file transfer, and access to bulletin boards and databases. But it is increasingly being asked to support other applications such as videophone, multimedia, and other commerical infomerical types of uses. Several tools exist to facilitate the otherwise complex and unwielding UNIX environment native to the Internet:

- E-mail.
- FTP (File Transfer Protocol)—lets client programs transfer files to/from servers.
- WWW (World Wide Web)/Mosaic—a graphical hypertext-based multimedia accessing tool.
- USENET—an e-mail conferencing system.
- Gopher—a navigation tool that facilitates search and retrieval of files from servers and databases across the Internet.

There is no formal management of the Internet per se. It's a cooperative effort among the divergent networks that make up its virtual form. However, there is an Internet Architecture Board (IAB) that was founded in the early 1980s by ARPA and has since become part of the Internet Society (a fund-raising organization). It's an architectural, not operations, group of a dozen people who oversee 50 subcommittees. Chief among these is the Internet Engineering Task Force (IETF), which takes most of the responsibility for Internet's technical direction, including much protocol and specification work on new communications software and revision of old systems.

Users access the Internet in two main ways: either with a single-user IP connection, using a 9600- to 14.4-Kbps modem on a dial-up basis to connect with an Internet service provider, or via a LAN/Router/28K to 64K leased-line connection to a service provider as part of a larger group. These service providers are either a commercial or a university/governmental entity. Commercial service providers typically charge a setup fee of $50 to $100 and $30 per month, plus a usage fee, or a $120 per month flat fee. These costs do not include any additional telecommunications fee for access to the service provider. Service for a multiparty LAN customer ranges from $300 to $1500 for a setup fee, and $200 to $600+ for a monthly fee and telecommunications costs.

Establishing an actual Internet presence on a backbone costs tens of thousands of dollars for the initial equipment, similar amounts on an annual basis for the system's administration staff, and telecommunications costs for ongoing operation of the system.

How does Internet work?

Users access the Internet via a LAN, a 9600- to 14.4-Kbps modem via the telephone network, or a dedicated higher-rate circuit such as switched 56 or ISDN connection to their Internet service provider's router. Let's say we want to send a large file across the network. Our file is broken into IP datagrams; each is assigned a destination address and then loaded into network packets with their own network addresses that might or might not be the same as the destination address. They travel around our office LAN to a router, at which point the network packet envelope is removed and the router determines the next destination on route to our final endpoint. It then advances the packet with the appropriate address to the next router. This continues as our message travels from our home LAN through a series of LANs to one with access to the public telephone network. Here, via a 9600-bit-per-second modem, it travels over the voice telephone network to the local universities' network. On their LAN, it travels to the regional WAN network, which uses T1 (1.5-Mbps) circuits and routers to link local schools to the state universities' LAN and Internet ANSnet backbone access. Once on the ANSnet, it travels over leased T3's (45-Mbps) to the midwest, where it joins a multistate regional WAN and travels to a state universities' LAN near our destination. From there it finds another telephone network modem through which it connects to our destination at 9600 bps.

As our file took this journey, it encountered many dangers, such as routing problems, exhaustion of IP addresses, blocking, delay, dropped bits, and voice-grade networks with their noisy, slow lines. Let's examine each of these further. With only a two-layer routing hierarchy, each router must manage, process, store, and update a tremendous amount of routing information to allow it to send packets on toward the correct destination. If the router does not recognize the destination, it will merely forward the payload on to another router in hopes that it will recognize the destination. This can go on endlessly with no hope of moving the payload closer to its destination. To address this situation, many routers will recall that a message has come to it in vain before and dump the message. This might occur with messages that have valid destinations that are merely not known to machines in a given area.

There is a saying that "it grew and grew until it became gruesome!" This applies to the Internet. As hundreds of new networks are added each year, the amount of signaling traffic, as well as the complexity of managing such a network, will eventually overwhelm the Internet, resulting in blocking, delays, and dropped messages. There is considerable, valid concern that voice-grade circuits present another danger. The voice world is hostile to data; noisy circuits and slow modems are an extremely inefficient way to communicate data. With 15,000 networks and more than 2.2 million users, Internet is in danger of running out of IP addresses. The two most discussed answers are to expand the address field or reuse host addresses. There is no easy solution, and many estimate the exhaustion of the IP addresses within the decade.

Strengths and weaknesses of the Internet

The Internet faces many challenges in its changing world. It has experienced exponential growth since 1983, with an average 10% increase in the number of users per month. It has grown from 562 terminals in 1983 to more than 2.2 million computers in 1994. At its current growth rate, it doubles every 10 months! This growth has come at a cost. It has dramatically changed the Internet's nature and application. It must now face dramatically different requirements than those for which it was originally designed. As mentioned earlier, the Internet was originally an R&D network for the military industry. It was subsequently expanded by NSF for academia. These users enjoyed a free and open environment and policed themselves—enforcing established norms of behavior and use of the Internet. With the explosion of commercial Internet users, new requirements have and will continue to be placed on the Internet as businesses attempt to establish electronic store fronts. Conducting business

via the Internet will place new demands for which
equipped to support:

- Security. For banking, financial, and credit card informatio
- Survivability. It's no longer an academic toy. New uses require
  time operability.
- Privacy. Currently the Internet is an open environment. Anyone can
  read anyone's e-mail, as they are forwarded through the network.
- Capacity. The network is quite vulnerable as type, frequency, and
  duration of usage increases.

There are also a whole new family of applications beginning to
emerge: multimedia, videoconferencing, voice, etc. The requirements of
these applications are very different than those of data. They are delay
sensitive, real time, connection oriented, and tend to have long holding
times. Internet, as a connectionless packet network, is incapable of supporting such applications.

Free and open no more, the Internet has become cluttered with millions of users from different backgrounds, with disparate and sometimes
conflicting views of appropriate Internet usage. This was poignantly illustrated by a law firm that sent unsolicited advertisements to over 5000
Internet new groups. In return, they received 30,000 angry e-mail messages (or flames in the Internet vernacular), some of which included death
threats. The mail-bombing programs, set up by angry users, repeatedly
crashed Internet in that part of the country. It cannot handle episodes
such as this. These types of abuses will eventually destroy Internet, which
lacks effective management and control to stop such activity. This illus-
trates the limits of a store-and-forward-type packet network. While adequate at meeting the needs of the original R&D community, Internet's
architecture and structure are incapable of effectively and efficiently supporting the requirements and needs of its new environment. It's only a
matter of time before an unresolvable disaster develops.

Conclusion

Internet's tremendous growth and popularity have clearly demonstrated
the need for a public data network (PDN). As we consider data's complex
and challenging requirements and speculate on the type of network required to achieve them, we recognize that it must be the right network
with the right capabilities to enable the continued dynamic growth of future voice, data, and video services. The alternative is to forever react to
new user needs and growth with dramatic costly changes to an ill-suited
network, in a futile attempt to meet expanding demands. Internet is a
store-and-forward network that was very good at meeting the requirements of the task for which it was originally designed—to provide data

user group of R and D scientists and
ıc arenas.

new group of users and requirements.
, which makes simplex routing and net-
difficult. Its new commercial users have
at might be hostile to the Internet and its
v have a wide range of new application re-
e transfer of data. We see a growing use of
.l as the need for real-time videoconferencing
app̤   -and-forward network is not designed to adequately sṳ   ̤ernet showed the need for a public data network, but is not ṳ   ̤rk capable of meeting that need. What is needed
is a family of robust, ̤ecure, survivable, switched public narrowband,
wideband, and broadband networks that support both circuit and packet
switching and will later support SONET, ATM, and STM in the broadband
arena. In addition, they must provide a host of voice, data, and video enhanced features and services beyond mere transport.

Wireless communications services

J.G. Hemmady
AT&T Bell Laboratories

Wireless communications have been in use in the world for many years.
However, it's only within the last decade that the radio technology is
touching the lives of individuals and is having an impact at a personal level.
The recent thrust of wireless technology is on the mobility market, including cordless telephones, paging services, dispatch/mobile radio, cellular
services, emerging satellite applications, and finally, personal communications services (PCS). This article discusses wireless thrusts on mobility
markets with special emphasis on PCS.

Personal communications services (PCS)

PCS are a set of capabilities that allows terminal mobility, personal mobility, and service mobility. *Terminal mobility* is the ability of a terminal (i.e.,
a telephone) to access telecommunication services from different locations while in motion, and the ability of the network to identify and locate
that terminal. *Personal mobility* is the ability of a wireless user to access
telecommunication services at any terminal on the basis of a personal identifier, and the ability of the network to provide those services according to
the user's individual service profile. The concept is part of such initiatives
as universal personal telecommunications (UPT) but is not currently deployed to any significant extent. *Service mobility* is an evolutionary concept that is the ability to use vertical features (e.g., local area signaling
services, Centrex, etc.) from remote locations or while in motion. PCS is

generally viewed as being significantly lower in cost than today's cellular service, providing significantly more mobility than today's landline service, and having voice quality and feature sets similar to today's landline service.

As personal communication services (PCS) evolve over the next decade, switching technology, both hardware and software, must evolve to provide personal communication networks (PCN) with seamless coverage, uniform functionality, and the ability to interpret the unique identity of each user and deliver the subscribed services. In the United States, the PCN will be an integration of the landline, cellular, and new wireless service provider networks (e.g., cable television) to create a single seamless network. Technology, federal regulatory environment, standards, and end-user and service provider market needs will govern the way the PCNs will evolve in the United States. The cellular and landline networks, which are two major networks in the United States, will evolve in supporting the goals of PCN.

PCS environment in the United States

Several factors govern the introduction and growth of PCS and PCN: technology, regulation, standards, and market conditions are among the most prominent. These factors associated with the environment in the United States are discussed in the following.

Technology

Technological advances are taking place at an unprecedented rate on all fronts—microelectronics, displays, radio, switching, intelligent networks, and signaling. Electronic advances continue to lower cost, size, and power levels of handsets, and combined with advances in battery technology, are increasing the talk and stand-by time of the portables. Advances in flat screen display technology and pen-based user interfaces are promoting personal digital assistants (PDA) and a variety of other portable terminals. Digital radio and microcell technology have significantly improved the spectral efficiency and have improved the in-building coverage and overall quality of service. In the near future, the Smart-card technology combined with Advanced Intelligent Network (AIN) and ubiquity of SS7 and ISDN will allow personal numbers, personal profile access, and "seamless" operation across multiple networks.

Regulatory environment in the United States The Federal Communications Commission (FCC) controls the allocation of spectrum for broadband PCS in the United States. The FCC has defined rules for auctioning the spectrum to potential PCS service providers (PSPs), and has considerably relaxed deployment schedules. These new rulings have received a lot of support and praise by the industry. All PCS licenses are in 1850-MHz–1990-MHz band. A 20-MHz spectrum in the 1910–1930-MHz band is set aside for unlicensed PCS services, a reduction of 20 MHz from the FCC's previous plan. A PCS service provider can bid for a 30- or 10-

MHz spectrum, and having won a 30-MHz allocation, is expected to provide 33% coverage in 5 years, and 67% in 10 years. The winner of a 10-MHz license is expected to provide 25% coverage in 5 years. The FCC has also announced rules for the existing cellular companies to play in the PCS market, primarily limiting the additional "in-region" spectrum they are allowed to license. A certain number of licenses are set aside for the minority, women-owned, and small businesses.

PCS standards In specifying the functionality of landline and cellular networks, standards bodies such as International Telecommunications Union (ITU), the Telecommunications Industry Association (TIA), American National Standards Institute (ANSI)[1], and the European Telecommunications Standards Institute (ETSI) must define standards for user services that provide seamless operation between landline and wireless networks.

Currently in the United States, the T1P1 committee has specified a reference model for PCS and has identified three different scenarios as potential ways of implementing the reference model[2]. The TR-46 subcommittee of TIA has also specified a reference model from a cellular service provider perspective[3]. Bellcore has identified a PCS wholesale access scheme that uses standard and open interfaces to the LEC Intelligent Network infrastructure[4]. Other organizations such as PCIA also have specified PCS reference models and proposed implementation options[5]. PCS data standards are not available at this time. However, specifications such as cellular digital packet data (CDPD)[6], which allows cellular users to send and receive electronic mail, facsimiles, and files over the airwaves, will set the tone for these standards.

Market drivers There are many market forces driving the development and deployment of PCS: end users, PCS service providers, and equipment vendors are the three prime drivers. Residential users seek a minimum-functionality wireless service with a price comparable to today's landline service. Business users seek a service similar to today's landline service and the convenience of a cordless phone with an extended range. Yet others seek a service that allows them to customize and obtain subscribed services from any phone at any location. All these needs have one central theme: to economically communicate at any time and from anywhere. PCS service providers will use their unique strengths to provide a competitive PCS offering. These companies include local and interexchange carriers, cellular providers, CATV operators, CAPs, satellite, paging, and special mobile radio services providers. Finally, the equipment vendors will drive the market by introducing low-cost, feature-rich, and spectrally efficient products.

PCS functional architecture

The evolution of landline and cellular networks in support of PCS is addressed by studying two views of the system architecture. A nodal archi-

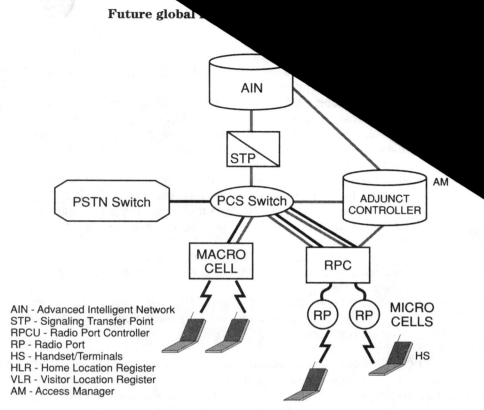

Fig. A-2. PCS functional architecture.

tecture view, which centers around a single "wirecenter" or a single switching system, and a network architecture view, which covers the interconnections of several nodes with seamless interworking.

Nodal PCS architecture Figure A-2 shows major functional blocks or network components in an end-to-end PCS system node:

- Wireless handsets provide single/dual-mode operation, depending on the air interfaces, and are lightweight, low-power, low-cost devices. Other wireless customer equipment includes PCs, PDA, facsimile, etc.
- Radio ports (RP) provide an antenna and optionally provide amplifiers, combiners, filters, and vocoders. These might be physically combined with radio port controllers.
- Radio port controllers (RPC) provide control functions for a multitude of subtending RPs, and provide speech coding, interface to switching network, amplifiers/combiners/filters, and optionally provide the access manager functionality.

HLR
VLR/AM OR VLR

/een multiple RPCs and
intelligent network using

management (registration,
nd locate), radio operations
management, and VLR func-
,e in an adjunct to the switch or

oal title translation and routing of
work elements using SS7.
) database, located in the network
permanent storage of subscriber-re-

- Visi.   /LR) database, collocated with the access manా   emporary storage of subscriber-related
  information.
- Personal number datau..se for personal mobility (PM), shown collocated with the HLR, provides routing information for terminating
  personal number calls.

PCS network architecture The Network architecture shown in
Fig. A-3 shows how various nodal systems —landline or cellular—in a
given region are interconnected using the signaling and AIN networks.
Additional capabilities for this architecture beyond those provided by a
nodal architecture are: signaling connectivity, query-response, and call-
routing capability among network elements. The query-response is used
to obtain subscriber information from the home location (HLR) for a roaming subscriber or for delivering a PCS call based on a personal number.
There could be one of more AMs in a region, depending on the number of
RPCs and traffic on the network. Typically, a given access manager serves
a given set of RPCs in a particular service area.

Radio technology Wireless media can be considered an avenue between the end user and the network. The elements that facilitate the access to the wireless medium are the end-user terminals (hand sets) and
the radio ports (base stations). These elements provide the network services to the end user as accommodated by the air interface employed. The
wireless communications link is rather fragile because of multipath propa-
gation of the signal, time-varying nature of the channel, and by the mobility requirements of the air interface employed.

The current cellular systems in the United States use both analog
(AMPS-FM) and digital (AMPS-TDMA and CDMA) air interface technologies. All of those systems operate at the 800-MHz frequency band and have
been optimized for voice traffic. The digital air interfaces operate in dual
mode, accommodating analog FM as well. The PCS operation will be at 2

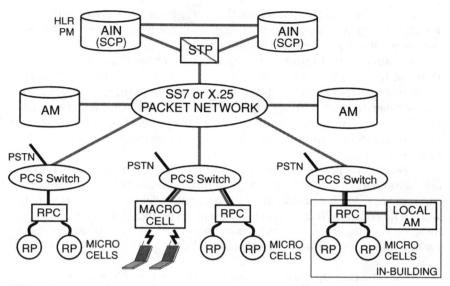

AIN - Advanced Intelligent Network
SCP - Service Control Point
STP - Signaling Transfer Point
RPC - Radio Port Controller
RP - Radio Port
HLR - Home Location Register
PM - Personal Mobility Mgt.
AM - Access Manager
PSTN - Publ. Swt. Tran. Net.

Fig. A-3. PCS network architecture.

GHz using the new spectrum allocated recently by FCC and will employ
digital air interfaces only. Digital technologies are being preferred over
analog technologies for the following reasons: higher capacity, better and
consistent quality, increased security, new network services, compatibility
with digital network, smaller equipment size, and soft handoffs (CDMA).

Currently, several air interface technologies are being considered in the
U.S. standards bodies (such as the Joint Technical Committee, JTC) for
eventual product offering. The current proposals at JTC encompass both
cellular-type systems such as modified forms of the current North American
cellular standard IS-54 (TDMA), modified forms of IS-95 (CDMA), modified
forms of the European standard (GSM) called DCS-1800, and noncellular-
type systems such as Wireless Access Communications Systems (WACS) by
Bellcore, Japanese standard Personal Handyphone System (PHS), a combination of WACS and PHS called Personal Access Communications System
(PACS), a hybrid TDMA/CDMA scheme by Omnipoint, etc. The cellular-
type systems provide higher capacity with low-bit-rate predictive coding

schemes (8 to 16 Kbps), whereas the noncellular type systems employ waveform coding schemes such as ADPCM at 32 Kbps to provide better-quality voice at the expense of reduced capacity.

Advanced intelligent network (AIN) Architectures that support PCS will require the networkwide signaling and the use of network databases in order to offer ubiquitous coverage and uniform service delivery. The AIN service control point (SCP) is a natural platform for the support of PCS. The following sections expand on four aspects of any PCS architecture and the natural role that AIN will serve in supporting the full deployment of PCS[7].

Centralized profiles database Secure user profile creation and storage is fundamental to PCS delivery. Centralized profiles will assure that services are delivered uniformly to the user wherever the user is in the network. The centralized user data includes user routing preferences, feature information, validation, and authentication information. Centralized databases support the following five aspects of PCS:

- Automatic registration: The user's particular profile is updated with the user's location and authentication information as the user moves between different coverage areas.
- Personal mobility management (PMM): PMM services are built on the centralized subscriber profile database architecture element.
- Provisioning: User and terminal provisioning can now be done at one point in the network, independent of the subscriber's point of service access.
- Fraud detection: Centralized profiles allow the monitoring of user access and usage pattern in real time.
- Subscriber profile manipulation: This allows users to interact with their individual profiles in real time and set up call screening lists for incoming calls.

Centralized routing database The AIN platform can provide the real-time routing intelligence required for PCS by receiving IN queries from the AIN call model and translating the received digits into the digits needed for routing PCS calls through the network. The AIN platform keeps track of the user's location in order for the network to always be ready to route incoming calls to the PCS user.

Feature-rich execution platform The AIN provides a robust platform for executing the service logic for calls ranging from simple POTS-like calls to the full complement of user-specific features. In addition, the AIN platform, through the service creation environment, offers service vendors the ability to rapidly create and deploy these services in an entire coverage area.

Standard interfaces Ubiquitous service is dependent on standard network signaling. The AIN platform products are based on standard SS7 interfaces, the TIA/EIA's IS-41 messaging standard.

PCS operations support The standards for operations, administration, maintenance, and provisioning (OAM&P) are being defined by the ANSI accredited T1M1/T1P1 ad hoc group on PCS[8]. The PCIA Technical and Engineering Subcommittee has also produced documents describing an approach for OAM&P.

The goals of OAM&P for PCS are to minimize the costs by making minimal additions to existing distribution networks, and by driving the functionality as close as economically feasible to the network elements (NE) that need to be managed. PCS management must allow multiple PSPs, multiple network management systems, and multiple vendors. For effective and efficient interworking, different management systems must share a common view of the network. Five major categories of OAM&P functions identified in IS7482 and CCITT Rec. M.3010 are: performance management, fault management, configuration management, security management, and accounting management.

Cellular evolution to PCS

Today, most 800-MHz cellular systems in the market use proprietary interfaces between the various network, components shown in Fig. A-2. These systems interconnect to the PSTN via trunks (either SS7 or per-trunk signaling) and do not provide interfaces to the intelligent network, e.g., service control point. By prior agreements, these networks allow roaming an area served by other service providers using IS-41 standard intersystem protocol.

These cellular service providers will offer two tiers of wireless service offerings in PCS domain:

- Low-tier offering: A low-tier service provides low-speed mobility, zonal coverage service at a much lower cost than current cellular.
- High-tier offering: A high-tier service provides ubiquitous, high-speed mobility at a cost comparable with current cellular.

Interworking between low- and high-tier services raises many issues, e.g., dual-mode handset cost, hand-off scenarios between low- and high-tier systems, tariff and billing schemes, etc., which are beyond the scope of this paper.

This two-tier strategy allows a cellular service provider to concentrate on the high-mobility market and maximize the reuse of the existing cellular network. Cellular service providers will approach the PCS market to provide low- and high-tier services in the following three ways:

A. Leverage and enhance their existing 800-MHz networks by adding low-tier PCS services capabilities, microcellular radio equipment, AIN, etc.
B. Expand the system capacity by obtaining an additional 10-MHz spectrum in 1.8-GHz range in existing service area;

C. Build a new 1.8-GHz PCS network in a different (out-of-territory) service area.

Each of these strategies places different demands on the cellular networks resulting in different evolution scenarios. However, there are many common elements between 800-MHz and 1.8-GHz systems as discussed below.

Cellular network enhancements An architecture for a low- and high-tier offering is shown in Fig. A-4. In case A above, cellular systems will leverage the benefits of the existing 800-MHz architecture while providing the necessary enhancements to meet the market needs at 1.8 GHz. The access manager and PCS switch in cellular platforms (e.g., AT&T's AUTOPLEX™ platform) could support both 800-MHz and 1.8-GHz radio port equipment[10], and fully support multisystem networking using IS-41. A major addition to the current cellular networks to provide PCS is advanced intelligent network (see Fig. A-4) to provide HLR/VLR and personal number databases for call delivery.

In cases B and C above, cellular service providers will need new radio equipment that operates at the 1.8-GHz frequency. However, the rest of the network elements (e.g., AT&T's access manager, cellular switch, etc.)

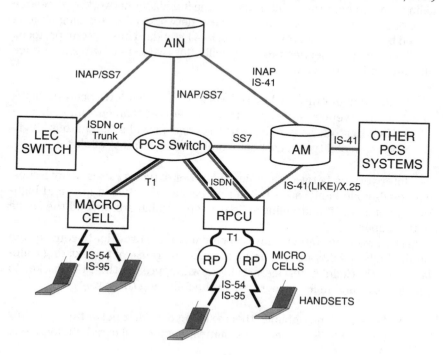

Fig. A-4. Cellular network evolution.

could be reused with relatively minor changes in their interfaces. To offer new "vertical" custom-calling features, cellular service providers will need software upgrades for the access manager and/or cellular switch. The cellular service providers will be able to wholesale excess capacity in their networks, i.e., allow other PCS service providers to hook up their radio equipment and AM to the cellular switch, when they have open interfaces.

ATM evolution   Future PCS platform will be designed to allow for graceful integration of expected developments in the landline network, in particular, ATM, SONET, and broadband technologies. Figure A-5 shows a PCS architecture that uses ATM switching for interconnecting multiple PCS switching centers in a particular service area. However, on the integration of ATM for this interoffice traffic, deployment of low-speed ATM access (rates less than 150 Mbps) enables further facility cost reductions in the radio port distribution network and the PCS switching centers. In this network the ATM connectivity, originally only connecting PCS switching offices, is extended to provide a broadband ATM-based transmission network for interconnecting radio ports and PCS switching centers.

In summary, the evolution of cellular network to PCS depends on the strategy cellular service providers implement. If the cellular service provider chooses to compete head to head with new low-mobility entrants, then the

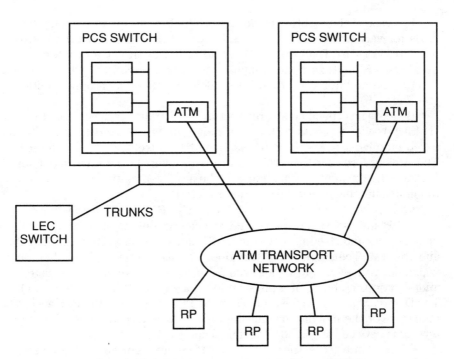

Fig. A-5. Cellular network evolution to ATM.

cellular network must evolve to include microcells, radio ports, radio port controller units, and possibly a new or modified access manager. If the cellular service provider chooses to use the new spectrum to increase the cellular network capacity, then the cellular network must evolve to include the up-banded radios and any new software for "vertical" custom-calling features. Either way, the cellular service provider will leverage and evolve the existing network to provide competitive service. Cellular networks will evolve in the future to incorporate ATM technology being deployed in landline networks.

Landline evolution to PCS

Landline networks owned by LECs can evolve to provide PCS by adding the wireless access, and by upgrading the switch and intelligent network software to provide the functionality needed to support mobility aspects of PCS. LECs can participate in PCS markets in two ways:

- As a retail PCS service provider. For example, they would own the spectrum in a region (in-territory or out-of-territory),
- As a wholesaler of their existing infrastructure to other PSPs who hold the spectrum licenses in their region.

For case A above, LECs could use proprietary or open network interfaces for various network elements because they would own the entire network. For case B above, LECs would prefer open network interfaces because a PSP in their region might own one or several PCS network elements. The architecture proposed by Bellcore offers such opportunities (see Fig. A-6).

Bellcore has specified an architecture [3, 9, 11, 12] that uses National ISDN (N-ISDN) switch access to the new radio equipment, and uses of SS7-based intelligent network protocols for radio control and PCS databases. The driving force for this architecture is to leverage the investment for central office switching, signaling, and the intelligent network, with a view to minimizing the investment needed by a PSP.

Four primary interfaces are defined: The P or port interface is between RP and RPCU and used by PSPs who own only radio access. The C or controller interface is between RPCU and the PCS switch and RPCU and the AM. There are many variations of this interface, depending on which network element provides the features (switch or access manager) and also on the choice of a physical interface (BRI, PRI, TR303, or T1). The D or database interface is between the PCS switch and the PCS access manager (or the mobility management database). Lastly, the A or air interface is between the handset/terminal and the RP.

Of these interfaces, the C interface has received considerable industry attention, and equipment vendors have started building ISDN-based prod-

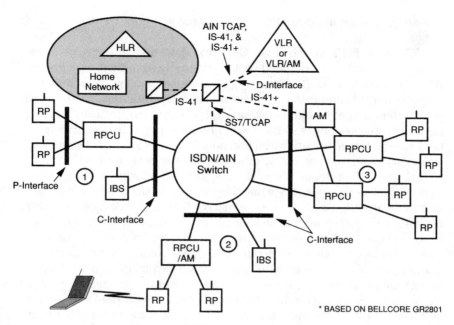

Fig. A-6. LEC network evolution.

ucts. The trend is to bundle RP, RPCU, and AM products in a PCS offering because of the proprietary nature of interfaces between these elements.

The LEC architecture in Fig. A-6 shows five major components: PCS switch, RPC, RP, AM/VLR, and HLR. A generic definition of PCS functions provided by these elements was given earlier in this appendix.

The local switch provides the N-ISDN interface (e.g., BRI) to RPC and an SS7 interface to AIN using AIN service switching point (SSP) triggering capability. The RPC uses a generic C interface, which consists of multiplexed N-ISDN BRI links to the switch. (Currently, a limited set of features is available on ISDN switches. Therefore, PRI is considered as a viable, longer-term alternative.) The RPC terminates N-ISDN signaling links (D channel), and under control of the AM, supports registration and call control, manages the interconnection of B-channel information to correct radio port link, and manages the assignment of B-channel information to air interface frequency and channel. The AM functionality has three optional locations: within the intelligent network SCP platform, within an RPCU, or centralized within the radio access side of the network.

Other wireless communications

Discussion of wireless communications will not be complete without other important mobility thrusts of wireless technology, including cordless telephones, paging services, dispatch/mobile radio, cellular services, and emerging satellite applications. The following is a brief treatise on these important thrusts.

Wireless PBX or wireless Centrex systems   In today's business environment, subscribers spend a lot of time away from their desk phones, which results in missed business opportunities and lower productivity. Wireless Centrex and PBXs are penetrating these traditionally wired business systems to help avoid this situation. These systems "logically" connect portable wireless phones to their wired phones so that incoming calls can be delivered to one or both wired and portable phones. The subscriber has the ability to answer the call at or away from the desk, thus reducing the "telephone tag" problem. Some of these systems provide wireless extension of wireline services only while they are within the business premises, and there are other systems on the market that allow seamless operation between indoor systems and cellular systems in the public domain. A specially designed adjunct to the customer-premises equipment, combined with a handset designed to operate in both environments, makes this operation possible. These systems also provide flexibility in providing the proper billing treatment for wireless usage within the premises, similar to the wireline Centrex systems. These systems are being built to work in both licensed and unlicensed bands, and are touted as a first step in the implementation of PCS.

Cordless telephones, Telepoint, and PHS   A cordless telephone is a low-power, low-range phone that is only associated with its own base station and provides two-way communication. The mobility is limited to the range of a single base station. A cordless telephone CT-0 unofficially refers to the analog VHF cordless telephones popular in the United States, and its first generation successor CT-1 is an analog UHF cordless telephone that is marketed but not widely accepted in Europe. Currently, digital cordless phones, using spread-spectrum technologies, are on the market. They provide a capability to choose from ten channels and provide a range of up to 1000 feet from the base station. These phones operate under FCC's Part 15 rule in the ISM band (46, 49, or 902–928 MHz).

Telepoint is a generic term for a form of PCS that provides cordless pay phone service to customer-owned handsets within limited range (about 100 meters) of base stations in public places. The telepoint service is expected to be priced below other cellular and PCS, and slightly above the cost of coin phone calls because the cost to service providers of installing telepoint base stations only in selected locations is far less than covering an entire populated area with base stations. To use the service, a subscriber, who is identified by the identification number of the handset, walks or drives within the range of the base station. The subscribers cannot receive calls because the initial telepoint systems are not capable of "locating" subscribers for call delivery. Some telepoint systems overcome this shortcoming by offering a paging option so that subscribers can be alerted to incoming calls over a wide-area paging network. They might choose to return the call by being within the range of a telepoint base station or by other means. A principal standard for telepoint called CT-2 (cordless telephone, second generation) uses a digital FDMA architecture. CT-2 products are available from European, Canadian and U.S. manufacturers but have not taken hold in the United States due to their limited capabilities.

Packet handyphone system (PHS) is a Japanese version of a cordless telephone. Its base stations use an ISDN interface to the central office equipment, and the air interface is designed to handle calls from a PHS terminal with a range of 100 to 200 meters. These systems claim low equipment and operational cost in comparison with cellular alternatives because of low power design of base stations and a small number of channels per base station. These systems are also designed for airport, shopping mall, or downtown-type applications where subscribers move around at walking speeds.

Wireless local area networks A motivating factor behind the use of wireless LANs is the reduction of the cost of adding, moving, and changing computer network connections within business premises. Until recently, the wireless data communications systems have been low-speed products due to a limited spectrum allocation. In 1985, the U.S. FCC allocated the ISM 2400-MHz band for wireless LAN applications, which has stimulated development of high-speed products (2 Mbps) like NCR's WaveLAN™. The current wireless LAN products, like the wireless PBX or Centrex products, operate in licensed as well as unlicensed bands (ISM bands 902–928, 2400–2483, 5725–5875 MHz). There is also progress on standardizing protocols, as evidenced by the IEEE 802 standards effort in specifying wireline and wireless interoperability standards, and "spectrum etiquettes" for future products to avoid interference with other wireless systems (e.g., wireless Centrex/PBX, monitoring devices) operating in the same spectrum. Also addressed by the standards bodies are concerns relating to privacy and security issues. The use of spread-spectrum air interface technologies have permitted 2-Mbps wireless LANs, which otherwise would have been limited to 300 Kbps[13].

Wireless data communications and CDPD Wireless data services include one-way broadcast services, electronic mail to roamers, two-way interactive services, messaging, public database access, etc. Demand for wireless data services is expected to grow rapidly in this decade, as evidenced by growth in many networks in the public domain (e.g., Ardis, Ram Mobile Data, etc.). Several techniques are used for wireless data communications.

- The packet radio allows a mobile terminal to send packets to the base station in its range, which in turn relays it to the data network for routing packets to final destination. The specialized mobile radio (SMR) uses the centralized packet radio technique, which is most common among service providers.

- Circuit cellular transfers data over a voice circuit between the wireless terminal with modem and the remote station with modem.
- Packet cellular is a new enhancement of cellular that allows packet data transfer. The cellular digital packet data (CDPD), discussed in some detail later in this chapter, uses packet cellular technique.
- The satellite technique uses a satellite network with other wireless or wired networks for data communications.
- The paging technique provides one-way data transfer for fast notification.

CDPD is a specification developed by a consortium of nine cellular carriers formed in 1992 and lead by IBM. It makes use of an existing cellular infrastructure and takes advantage of the idle time on cellular voice channels to fill with packet data. Interfaces at the cell sites and switching system side separate these packets and route them via an intermediary system to the final destination. The data rates of up to 19.2 Kbps are planned at this time. CDPD supports subscriber roaming through an automatic registration process, and also facilitates connection delivery and call handoffs during cell transfer. It provides enhanced security against both casual eavesdropping and fraud.

Summary

In this paper we have addressed how the current communications networks might evolve to support the vision of personal communications networks of the future. The goal is to create a single, seamless network to provide services associated with a mobile individual or a personal identifier, and not simply a terminal. Specifically addressed were the evolution of two most common networks in the United States—the landline and cellular networks. Several new technologies that facilitate the creation of a PCN—digital radio, signaling system 7, intelligent network, digital switching, and operations support technologies—were briefly discussed.

The cellular service providers will approach the PCS market by leveraging their existing networks to provide a combination of low- and high-speed mobility service or to use acquired spectrum to increase the capacity of the high-speed mobility services. Each of these two strategies will place different demands on the cellular-based PCS networks, resulting in two different evolution scenarios.

Landline networks owned by LECs will evolve to provide PCS by adding the radio access, and by upgrading the switch and intelligent network software to provide the other functionalities needed for PCS. LECs will participate in PCS markets as wholesalers of their existing infrastructure to other PSPs and/or as retail PCS service providers.

For completeness, a discussion of other important mobility thrusts of wireless technology is included. These are: cordless telephones, paging services, dispatch/mobile radio, cellular services, and emerging satellite applications.

Acknowledgments

The material on PCS is based on a recent paper presented by the author at Globecom '94[14]. The author acknowledges comments and contributions by Jorge Maymir, Dave Meyers, Brian Bolliger, S. Krishnamurthy, Nitin Shah, Anil Sawkar, Larry Gitten, and Lee Sneed.

References

1. ANSI will be the official "source" of all U.S. standards. T1S1, T1M1, and T1P1 are subcommittees of the standards committee T1, which is sponsored by ATIS, an alliance for telecommunications industry accredited by ANSI. The European Telecommunications Standards Institute (ETSI) must define standards for user services that provide seamless operation between landline and wireless networks.
2. A Technical Report on Network Capabilities, Architectures, and Interfaces for Personal Communications—T1P1/93-062R3, Nov 1993.
3. TR46—Draft PN-3168 Personal Communications Service Descriptions, Sept 17, 1993.
4. SR-TSV-002459 PCS Network Access Services to PCS Providers, Issue 2, Oct 1993.
5. Telocator now (PCIA)—PCS Standards Requirements Document—Service Description Standard, TE/92-07-09/104.
6. Cellular Digital Packet Data System Specification, Release 1.0, July 19, 1993, and AIN 0.1—advanced intelligent network (AIN) 0.1 Switching System Generic Requirements TR-NWT-001284, Issue 1, August 1992.
7. AIN 0.2—advanced intelligent network (AIN) 0.2 Switching System Generic Requirements GR-1298-CORE, Issue 1, November 1993.
8. T1M1.5/93-001R1—Proposed Draft Standards—OAM&P Interface Standards for PCS.
9. Generic Criteria for Version 0.1 Wireless Access Communications Systems (WACS) TR-INS-001313 Issue 1, October 1993, Revision 1, June 1994.
10. Interworking between low-tier and high-tier services is a goal of this network, which needs further definition.
11. Personal Communications Services (PCS) Network Access Services to PCS Providers SR-TSV-002459, Issue 2, October 1993.
12. Switching and Signaling Generic Requirements for Network Access Services to Personal Communications Services (PCS) Providers GR-2801-CORE, Issue 1, December 1993.
13. Bruce Tuch, "Development of WaveLAN™ and ISM Band Wireless LAN," AT&T Technical Journal, July/August 1993.

14. J. G. Hemmady, J. R. Maymir, D. J. Meyers, "Network Evolution to Support Personal Communications Services," Proceedings of Globecom 94.

To chart a successful course for narrowband, wideband, and broadband future offerings, a major trial took place in the early 1990s called COMPASS. Let's close our analysis by reviewing the outcome of this trial, which attempted to address several of the key technologies and market opportunities in terms of how the prospective users used and appreciated these new capabilities in their daily operations.

COMPASS points the way to ISDN's future

Peter Heldman and Jack Warner
Contributions by Eve Aretakis, Steve Dunning, and John Strickland, including interviews by Bob Stoffels, editor of *America's Network*, with Tom Madison, U S WEST, Dana Ziteck, AT&T Network Systems, Ken Hovaldt, Fujitsu Network Switching, Fred Fromm, Siemens Stromberg-Carlson, and Tom Bystrzycki, U S WEST.

On March 19, 1991, at SUPERCOMM in Houston, Tom Madison and Tom Bystrzycki of U S WEST announced that the Regional Bell Operating Company was "embarking on the nation's first comprehensive series of narrowband, wideband and broadband network application trials." Leading a supplier team that included AT&T, Fujitsu, and Siemens Stromberg-Carlson, U S WEST described its role as "charting a course to the design, standardization, provisioning, and operation of high-speed networks for the future." That effort—Communications Programs for Advanced Switched Services (COMPASS)—is nearing completion after three years. Looking toward the national information infrastructure, COMPASS's lessons will be valuable.

Goals

COMPASS started as part of a series of informal dialogues between U S WEST's new product/service thinkers and communication suppliers. From its inception, COMPASS took an uncharted course, revolutionizing product/services market assessments. Leading providers, suppliers, and users collectively explored the exciting new world of advanced communications. The program's intent: prove that advanced, wider-bandwidth switched communication services could be implemented sooner than believed possible, while exploring the end-user broadband service domain. The group sought to accomplish three goals:

- Assess future market needs and identify new applications.
- Assess new and emerging networking technologies, including frame relay, switched multimegabit data service (SMDS), asynchronous transfer mode (ATM), synchronous optical network (SONET), etc.

- Understand how to move from today's telephone network to tomorrow's narrowband, wideband, and broadband network, as existing and future market demands and opportunities move from POTS (plain old telephone service) to PANS (pretty awesome new services). See Fig. A-7.

Plans focused on projects that could start in first-quarter 1991 and extend about three years, when it seemed prototype broadband ISDN, with its 155-Mbps ATM interface, could be trialed with leading-edge end users. The challenge was to trial the complete spectrum of wider-bandwidth communications: narrowband to 1.5 Mbps, wideband 1.5 Mbps to 50 Mbps, and broadband 50 Mbps and beyond. Formulators of this effort were frustrated by the pace of standards development and made a deliberate effort in COMPASS planning not to let standards issues slow down the trials' schedule. Part of the intent was to let COMPASS trials establish the knowledge base for future standardized services and products.

Project design

Technical possibilities for wider-bandwidth communications present several options. Most industry dialogue focused on advocating one approach over another. COMPASS originators sought a different approach. They planned to put into trial service, with real end-users in real applications, an array of wider-bandwidth switched networks to learn end-users' evaluation of the new services' utility and, by implication, the underlying technology utility. Basic COMPASS objectives were:

- Test end-user reaction in real switched-network applications that were part of their regular operations;
- Focus on learning service, not product, "prove in";
- Use production or quality prototype equipment, available in the proper time frame;
- Maintain cooperative relationships among U S WEST, AT&T, Fujitsu, Siemens Stromberg-Carlson and others.
- Trial broadband ISDN switched 155-Mbps ATM interface applications. All suppliers would provide a B-ISDN switch.

From these objectives, five specific trial projects were formulated:

1. High-speed data networking using narrowband ISDN (N-ISDN).
2. Executive desktop video networking using 45-Mbps transport, eye-to-eye terminals; subsequently compressed to 384 Kbps.
3. Switched DS1 circuit service.
4. SMDS networking.
5. Broadband ISDN ATM networking.

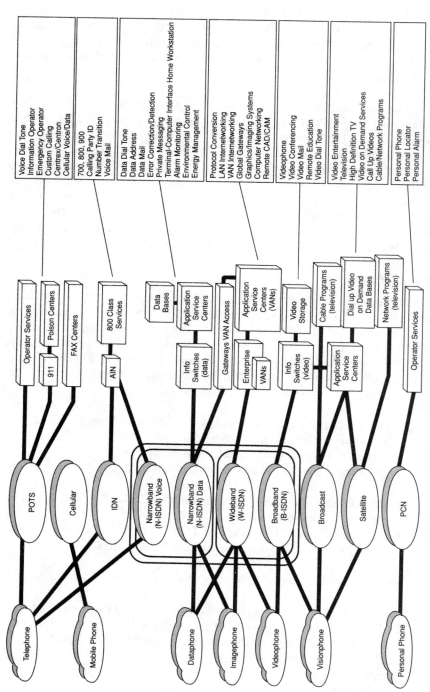

Fig. A-7. Information telecommunications vision of the future.

A learning experience

Bob Stoffels, Editor

"COMPASS was a starting point. It moved us way down the road. We're nowhere near where we need to be, but we're making significant moves forward." That's the assessment of Tom Madison, retired president of U S WEST Communications. "One day, we were having a discussion about the services customers expect," he says. "We then visited one of those customers—in this case, a hospital in Minneapolis—and we talked with the head radiologist. He did have need of a service, one that we could help provide. It had to do with the transmission of medical images among the radiologists. It was a matter of marrying applications and technology. And that was the beginning of COMPASS." Madison points out the need to get suppliers involved, preferably on a totally cooperative basis. These suppliers, along with U S WEST, would determine how customers view some existing services and build their needs into future products and services. "The bottom line was that we learned some interesting things," Madison says. "We learned some things that didn't work, and we learned some things that did work, and we got closer to the customers. As a matter of fact, as we got near the end of the trial, the customers didn't want us to take the equipment out."

Customers were COMPASS's foundation, Madison emphasizes. The project pulled in suppliers and service providers to satisfy users' needs. "Everyone wins," he says. "For instance, we used the eye-to-eye terminals for video conferencing. It was a phenomenal thing. It's the way of the future. And we learned that if you try to get by cheap, with bandwidth that is too low, you are not going to satisfy the customer." Madison notes the age-old problem of ubiquity: a service is more valuable if many people have it. He also acknowledges that many factors play a role in developing and providing services. "You have to realize that we have a great deal of embedded plant," he says by way of example. "How do we take advantage of that embedded plant and build some of the services around it?" How did customers react? "We got a lot of cooperation from a lot of customers. That was critical. Customers ranged from the small real estate company to the huge University of Minnesota operations. We had cooperation from everybody. As I said, we're not where we need to be, but we're way ahead of where we would be were it not for COMPASS."

Trial description, results

Project 1—High-speed data networking Project 1 focused on various aspects of N-ISDN data networking over three projects. The first demonstrated benefits of electronic document imaging over a wide area network (WAN). Ten University of Minnesota students were linked to an AS/400 host computer at the university. At home, students used a single ISDN BRI (basic rate interface) line for both high-speed switched data and digital voice services. The students' IBM PS/2 computers connected to the AS/400 host via one of the ISDN terminal's "B" channels and used packet-switched data service accessed through N-ISDN data terminal adapters. This arrangement allowed students to look up online text and scanned images, as well as perform collaborative computing remotely. The network supported multiple simultaneous applications sessions with the host, downloading large, full-screen images within 12 seconds. That was impossible or impractical with former modem connections over the analog voice telephone network.

The second trial project bundled two high-speed B channels for distance learning. Users at the University of Minnesota rhetoric department tested PC-based video conferencing and multimedia document-sharing applications over a 128-Kbps ISDN circuit-switched network. This provided flexible, high-speed access to classroom resources from individual homes, a crucial benefit to physically challenged students.

The third trial deployed real-estate imaging and multiple listing service (MLS) information over an ISDN B channel to remote real-estate agents. This application not only replaced conventional paper MLS books, but also provided online full-color photos of homes and neighborhoods. The network also offered more up-to-date sale listings and online access to more detailed information than conventional MLS books.

The following observations and conclusions resulted:

- A switched public network that supports common standards and interfaces is ideal. National ISDN standards were particularly useful.
- Ubiquity of service is essential for wider application and success. Considerable technical, provisioning, marketing and customer support is essential.
- N-ISDN equipment was operationally successful, with considerable end-user acceptance.
- N-ISDN had many applications, contrary to the common view.
- Loop plant requires individual verification of N-ISDN compatibility.
- N-ISDN pricing structures are crucial for customer use and acceptance; otherwise, users will remain on voice-grade modem over the public switched voice network or private leased lines.
- End-user applications require fully distributed N-ISDN connectivity and availability. Lack of this interconnectivity/availability is a major industry problem.

A living laboratory for technology

Bob Stoffels, Editor

"Siemens Stromberg-Carlson was involved with many of the COMPASS trials," says Fred Fromm, senior vice president. "Distance learning was important; so was medical imaging. More recently, we have supplied our ATM switch and are doing interoperability tests with other vendors."

Fromm explains how the ATM switch fit in the network: "Our vision of the ATM switch is that it can be applied as a standalone switch if the network employs an overlay topology. Or it can be done in an integrated way if the goal is to extend the existing network fabric. The idea is to give the network operators the flexibility to use it in the best way, based on the economics of the situation. For example, where a company is pressing hard for video, they will likely provide an overlay network. But in areas where they are trying to build a full-service network, it's likely best to build it on top of the existing structure."

Fromm says the various trials were valuable from two standpoints: they demonstrated that the technology could do the job and established price points that show what the customer will pay for the capability.

"The COMPASS trials amounted to a living laboratory to demonstrate the technology and test the market," he says. "We discovered we have to be much more involved with application trials than we would have 10 years ago. We have to be much more involved because the application has a great deal to do with the technology."

What applications seem important? "We're all struggling with the economics of the situation. And we're gravitating toward entertainment, where we think customers will be able to pay," Fromm says. "But hopefully, by doing this, we'll bring mass-market funding in, and this will allow us to add incremental services like medical services and do it economically. If you want to build this solely on business applications, someone is going to have to pay an awful lot of money. That's why we'll start with something that people are willing to pay for."

And the participants? "They got along extremely well," Fromm says. "We initially had those normal fears about revealing secrets and weaknesses to our competitors. But nothing came of it. U S WEST focused the applications in such a way that we weren't required to disclose any secrets. We worked together, and we showed ourselves that working together can be done in a manner that is not threatening."

Project 2—Executive desktop video networking In Project 2, U S WEST senior managers in five states tested a prototype eye-to-eye switched videophone network. A special overlay prototype network was provided. In phase 1, executives were linked via 45-Mbps inter-LATA and intra-LATA service.

In phase 2, this group tested the same system with data rate compressed to 384 Kbps. Calls were routed by prototype STM switches. These directed traffic on a dial-up basis, much as today's POTS network operates. Videophone terminals provided users true eye-to-eye contact. This was accomplished by optically folding their images to an internal camera via half-silvered mirrors. Video calls were established using narrowband ISDN phones. This "friendly" human interface allowed users to establish two-party and multiparty conference calls and to speed dial calls to participants, as users today make POTS calls. The multiparty videoconferencing capability supported split-screen viewing of conferees via a central bridge, collocated with each switch.

Observations and conclusions:

- There is a strong videophone market.
- Customers accepted this high-quality broadband system, since a "telepresence" was achieved.
- Videophone terminals and multimedia terminals are different devices for distinct applications, although the two might share some overlapping features and functionality. Videophones evolve from telephones; multimedia terminals evolve from PCs.
- Quality video demands quality audio. Full-duplex speakerphones address annoying clipping problems.
- Eye-to-eye capability is essential for achieving a telepresence in any videoconferencing application. With desktop applications such as videophone, the parallax problem becomes acute.
- Delay and latency are crucial problems that must be addressed to achieve usable quality. This problem becomes more acute with highly compressed video.
- Achieving lip sync by buffering audio signals to compensate for video processing delay only partly solves the problem; overall latency end-to-end remains a crucial issue that can and must be resolved with more efficient processors/algorithms or increased bandwidth.
- End-to-end delay requirements add a new dimension to network configuration and design, forcing network engineers to think beyond the line that divides network switching and transport from applications.
- Service availability must be ubiquitous.
- Reliability and availability are essential. Success breeds reliance on a system, so the system must not go down.
- Lighting remains a big issue; keeping subjects adequately front lit is challenging in a home office.

- The application must be easy to use, with little or no training. It helps if the CPE is an evolution of existing technologies with familiar human interfaces.
- Businesspeople want high-quality videophone (45 Mbps) with eye-to-eye capability to maximize communications and reduce travel.
- Users are sensitive to quality of the spoken word and video image, thus the problems with lighting, voice quality, and delay.
- Human-factors design for CPE control and call control are important for success.
- Video telephone service will require new installation and support skills from telephone company operations people.
- Business video telephone service must support document imaging and other vertical features beyond voice and video.
- Excellent-quality switched broadband communication can be provided now and is highly desirable; transport price is the major obstacle to widespread deployment.

The five COMPASS projects—and certainly the two presented here—confirm the value of real applications of wider-band services. Excellent acceptance was found for data networking using basic rate ISDN, wideband video networking, switched DS1, SMDS, and broadband ATM. There is a desire and need for wider-band communications applications—a need suppressed by the limitations of today's voice network. Transmission speed is not crucial. More important are practical issues related to communications network deployment, cost-effective equipment, and human-factor design.

Showing the potential

Bob Stoffels, Editor

"The most powerful thing that took place was the notion of getting the end user involved. This, and the vendors working with the service provider to craft offerings the customers said they could use."

That's the view of Dana Zitek, an AT&T Network Systems regional vice president. "In this industry," he continues, "the technologists often try to push what they have. COMPASS was different. One of our first steps was to find what would play in the marketplace."

AT&T, Siemens Stromberg-Carlson, and Fujitsu got along "very well," says Zitek. "I don't remember any friction points at all. U S WEST did a good job of managing that and was careful to maintain proprietary needs. It was a good example of how vendors can work together."

As to lessons of the trials, Zitek says it was more a matter of confirming things than learning, as with the eye-to-eye technology of the video conferencing trial. "We thought the eye-to-eye feature would be important. And it was."

Participants also learned much more about price points—what the customer would and would not buy. "In some cases, the customer is willing to pay more than we would have believed," Zitek says.

On cooperation: "I think one overall lesson was that we could have an open relationship and that it would work better than throwing a lot of requirements on the table and sifting through them and quoting. COMPASS was a partnership. I believe this is the way this industry can make progress."

"A number of the trials have resulted in products. We proved the need for SMDS in COMPASS, and AT&T has formed a whole business unit to offer it. And of course, we're working actively on the video project. We're targeting a real low-cost videophone."

"Then there was the ATM work," he says. "That is moving very rapidly and has demonstrated the need to transmit the right information down the pipe for broadband networks." COMPASS "determined the incredible potential for a wide variety of communications capabilities that could exist in our networks."

Showing a clearer path to the future

The COMPASS series of ISDN trials started through informal dealings between U S WEST new product/service thinkers and suppliers. The program demonstrated that advanced, wider-bandwidth, switched communication services could be implemented sooner than many believed; future market needs and applications would appear quickly.

COMPASS brought together leading providers, suppliers, and users. Many suppliers were involved; AT&T, Fujitsu and Siemens Stromberg-Carlson played the largest roles.

Project 3—Switched DS1 An adjunct was added to the AT&T 5ESS switch to facilitate DS1 switching. The trial consisted of three phases, each with a different customer. In the first phase, medical images were transferred from three locations to the main office of CDI, a large Twin Cities radiology clinic. In the second phase, the Minnesota Department of Transportation established video conferencing links among three sites; in the third phase, HealthOne medical center set up video conferencing links among several sites.

The DS1 switching platform consisted of a DACS IV, an applications processor, and 5ESS central office switch. The DACS IV provided hardware and software to route DS1 circuits. The 5ESS switch received the call setup/tear-down signal from users via an ISDN basic rate interface (BRI) and routed these signals to the applications processor, which supported call control, translations, etc., along with directing the DACS IV. Users established calls by dialing *T1 (*81) for call setup and *T9 (*89) for call tear-down.

Observations included:

- Human factors such as call-control procedure can make or break the application, even if it's cost effective;
- Both the employee performing the work and the business decision maker must be satisfied with the application.
- Both the technology and human interface must succeed to achieve a winning offering. Since this project's human factors failed to support applications, underlying network technology—no matter how flawless—never could become a viable offering.

Project 4—SMDS AT&T and Siemens Stromberg-Carlson switched multimegabit data service (SMDS) equipment was used in multiple trials. AT&T's BNS-2000 data switch platform was trialed in three phases. During phase 1, U S WEST conducted internal testing of interfaces between customer premises equipment (CPE) and the switch. In phase 2, the Westnet Consortium (an affiliation of western universities) and U S WEST Technologies addressed LAN-to-LAN interconnection applications, testing different mixes and types of traffic, network engineering, deployment, installation and maintenance of SMDS networks. In phase 3, CDI radiology clinic transferred medical images from its three remote sites back to the main office.

The best trial configuration in each demonstration was composed of an AT&T BNS 2000 SMDS node, supporting both DS1 and DS3 interfaces and a host of SMDS access classes. Customers connected to the BNS platform via dedicated T1 or T3 subscriber network interfaces (SNI). They accessed these circuits via SMDS terminal adapters, which converted the payload into SMDS format, adding signaling and address information. AT&T's Starkeeper OS platform provided administration provisioning and operation functionality for the BNS switch.

Siemens Stromberg-Carlson's EWSxpress 2100 asynchronous transfer mode (ATM) service access switch platform provided DS3 SMDS services between numerous 3M Co. sites and the Minnesota Supercomputer Center. The trial demonstrated the ease of provisioning and adding subscribers in a dynamic network environment. In addition, using DS3 SMDS, 3M could access remote sites on demand at higher speeds than possible with previous lower-speed leased-line network solutions.

Observations included:

- End users prefer switched solutions over private-line alternatives for data transport, if switched services are easy to use.
- LEC craft and salespeople must be trained extensively for successful SMDS deployment.
- Many applications could have used more than the 45-Mbps maximum SMDS access rate.

- The shared SMDS network solution offered customers tremendous cost savings versus existing point-to-point mesh networks. Paradoxically, current point-to-point T1 circuits are profitable to local networking providers.
- One or more interexchange carriers must support any switched data transport solution to be cost effective end to end.
- Switched data transport could cost end users less than private network solutions.

Designing for the human factor

Bob Stoffels, Editor

"We at Fujitsu got involved with COMPASS in the first year of the trials from the standpoint of narrowband ISDN. Later, we added video distance learning and participated in some of the medical trials." Ken Hovaldt, Fujitsu Network Switching's senior vice president, says the trial taught practical lessons about ISDN. "We had noticed that everybody talked about the technology," Hovaldt says, "but nobody had any proven track record or, for that matter, any application for broadband ISDN. One of the many things we learned was that applications have to fit within the customer's current environment. It's very hard to get people to do something totally different. That's the human-factor part of it." That dictated special equipment. "It was a situation where we actually had to build proprietary terminals," Hovaldt says. "It required a lot of study up front. I think we vendors have to understand that technology isn't just one giant leap," he continues. "It's really a gradual transition from what the customer is currently doing to what the customer wants to do."

Fujitsu started out small with each application. A few marketing people went along with operating company people to interview candidates. As appropriate applications were identified, they were turned over to systems engineers for development. For example, "We had developed a broadband switch and also some terminal equipment to demonstrate its throughput. One of the pieces of equipment was a video telephone. It ran at a variety of speeds, some as high as 100 Mbps. It was fine video, believe me. But nobody could afford it; it was just to demonstrate the kinds of things that could be done. Using this as a starting point, we were able to work with customers to develop applications they could use. Sometimes this resulted in actual product offerings; sometimes nothing came of it. The end user was not going to openly embrace new technology for new technology's sake," he emphasizes. "They have certain investments in their embedded base, and they have certain ways of doing business. Unless you can show them there is greater benefit from an economic point of view as well as from an efficiency point of view, they are not going to change. This was one of our most significant lessons." Hovaldt notes another positive outcome: the opportunity to perform interoperability testing with CPE and switch vendors. Fujitsu remains eager to continue working with vendors who work with open interfaces and test interoperability. "COMPASS has been a great help," he concludes.

Project 5—Broadband ISDN ATM networking Project 5 was composed of two customer trials. The first was a high-definition distance learning trial among teaching hospitals, including the University of Minnesota Hennepin County Medical Center and the VA hospital. The second was a high-speed computing application tested jointly among Honeywell, the University of Minnesota, and the Minneapolis Supercomputer Institute. These trials occurred in multiple phases as AT&T, Fujitsu, and Siemens Stromberg-Carlson in turn supported the trial with their ATM switches and cross connect equipment.

Interoperability issues also were explored. The computing trial featured interoperability testing between AT&T's prototype ATM switching platform and FORE System's ATM cross connect. Siemens Stromberg-Carlson's switch interoperated with Newbridge's Digital Link and Cisco ATM equipment used in the high-definition distance learning trial. U S WEST, Fujitsu, and Siemens Stromberg-Carlson plan further interoperability testing of ATM switching platforms. Customers accessed this overlay ATM trial network via generalized terminal adapters (GTAs), which supported a number of inputs (DS3, BRI, 802.6), converted them to ATM cell format, and provided addressing information.

Remote GTAs, located at customer premises, connected to the central office (CO)-based ATM switch platform via dedicated OC-3c 155-Mbps optical links. AT&T provided a prototype B-ISDN/ATM CO switch to support core switching. Fujitsu provided a FETEX-150 VI CO switch with OC-3 user-network interfaces (UNIs) and SMDS SNIs. Siemens Stromberg-Carlson's ATM switch platform supported a host of interfaces. The high-definition distance learning trial graphically illustrated the power of the single 155-Mbps ATM B-ISDN interface connecting each hospital. Students at all locations could choose among three monitors to view the instructor, students at other sites, or specific students asking or answering questions. They also could view high-resolution digitally-scanned 35mm slides, X-rays, or other information, documents from document cameras, and the instructor at a white board. The instructor could annotate material electronically using a remote-control cursor and could view and direct questions to any student.

In the manufacturing process control trial, synchronous optical network (SONET) OC-3c links connected CO-based ATM switches to high-speed workstations, supercomputers, and other terminal gear via TAXI 140-Mbps interface.

Observations included:

- CPE remains the crucial satisfier/dissatisfier for end users.
- Lighting and audio remain the biggest problems with two-way video communications.
- Human factors such as instructor and student control of media, camera tracking, "air mouse cursor" control, etc. became crucial. As the trial developed, users refined applications requirements.
- Developing and supporting CPE applications required sustained attention because of their complexity and evolving user requirements and expectations.
- Users' expectations for performance increased rapidly. As the trial progressed, they became intolerant of even minor CPE problems.
- ATM handles a dynamic traffic mix well and offers great flexibility, but as applications approach and exceed the great bandwidths available, hard engineering capacity rules might mitigate this flexibility.
- Much must be learned to engineer ATM networks efficiently as network designers struggle to quantify traffic parameters for the evolving mix of new services.
- ATM standards bodies have much work before achieving full interoperability among vendors' equipment. Until then, interoperability will be relegated to the lowest common denominator.
- The 155-Mbps B-ISDN interface has tremendous potential for telecommunications to substitute for physical presence because multimedia communications can be facilitated effectively (as in medical radiology distance learning).
- Converting today's media from its normal form to a useful digital form is a significant issue for medical images, photos, vu-graphs, movies and videos, among others. Time required for this conversion created significant dissatisfaction in the medical trial. Improvements are required to improve scanning time of hard-copy images into digital form for transport over an ATM network. User interface and CPE issues are crucial to the network's success.

COMPASS overall

COMPASS demonstrated the value and impact of wider-bandwidth, switched services. COMPASS also taught many lessons regarding overall network communications. In general, users want and need wider-bandwidth communications for data and video communications—applications suppressed by the current voice network's limitations. Speed is not crucial. N-ISDN, W-ISDN and B-ISDN applications all received favorable reviews. The issue is proper application of these baskets of network services to address end-user communication requirements. Key problems center on practical issues related to communications network deployment, cost-effective equipment, and human-factors design. Human factors deserve special attention, since all trials showed their importance. COMPASS also demonstrated a model of industry cooperation among firms working on the frontier of new service/product definition. The effort involved considerable innovation in a relatively short time, compared with slow standards-driven efforts that have dominated the industry. COMPASS provides a basis for future industry advances.

Not just a voice telephone company

Bob Stoffels, Editor

"When COMPASS was first launched, there were really two thrusts," recalls Tom Bystrzycki, executive vice president at U S WEST. "The first was to determine whether there were markets for the services we had in mind. That made it applications based, rather than technology based."

"At the time," he continues, "I was in charge of operating the network. So I was very interested in the second objective—determining whether technology was ready for these applications.

"On both counts, I think we proved a number of things to ourselves," he says. "One of the most significant things is that you have to design an infrastructure with all sorts of services in mind. Or, as I am fond of saying, don't sail the aircraft carrier up the river."

"So, while you might build an infrastructure and introduce entertainment services on it, you have to keep in mind multimedia services, data services, and the like," Bystrzycki says. "COMPASS influenced our thinking. It showed us that if you build the right infrastructure, there is this whole wealth of opportunities coming to you."

He offers an example. "If you build an infrastructure solely around teleradiology, you will find that the cost of the infrastructure so overwhelms the application that you can't afford to build it. On the other hand, if you build an infrastructure that can carry a whole complement of other services—including entertainment services—you will choose your architecture a little differently. Also, you'll find that the incremental cost of adding teleradiology to this infrastructure is rather small."

Another COMPASS lesson: all services need not focus on the maximum capability available. Clever things can be done with ISDN, such as stacking channels and detecting bit error. "In fact," Bystrzycki adds, "people who are working on our broadband program are now redefining it as a full spectrum of service offerings."

Why include coax in the network? "If you sat down with me and looked at where we were planning the nodes," he explained, "you wouldn't be troubled at the use of coax. First of all, I think it will disappear in rather short order. Right now, it's an economical choice. But later on, the economics of fiber will change, and then the coax will start to disappear—I would say within four years or so."

What happens next? "We can't end COMPASS," Bystrzycki says. "We can end the name, but we can't end doing the applications. We are finding that new applications are coming at us from all over. Education applications are a good case in point. They range from narrowband, where someone sits at a PC and gets assignments and information, all the way to teleteaching—which, of course, is a very wideband application. COMPASS demonstrated to our customers that we are not just a voice telephone company."

Peter Heldman is Senior Manager, Strategic Opportunities for AT&T. Jack Warner is Senior Bell Laboratories Field Representative for AT&T. John Strickland, U S WEST; Eve Aretakis, Siemens Stromberg-Carlson; and Steve Dunning, Fujitsu also contributed to this article.

Acknowledgments

The COMPASS Program was based on Robert K. Heldman's five books:

Telecommunications Management Planning: ISDN Networks, Products, and Services; *ISDN in the Marketplace*; *Global Telecommunications*; *Future Telecommunications*; and *Information Telecommunications*, all published by McGraw-Hill, Inc. (See Fig. A-8.)

Questions and answers

Challenge of the 1990s for the information marketplace

Question: RBOCs (LECs) have experienced a lot of changes in the 1980s and early 1990s. Will this continue throughout the late 1990s?

Answer: Yes, most RBOCs have consolidated their former BOCs and restructured them around market units. Now in the 1990s, these units must meet the needs of their customers.

Question: What will be the challenges of the 1990s?

Answer: First and foremost, the answer is services, services, services. As we become more in tune with the marketplace, we must provide not only voice services, but also data and video services as well for both the public and private arenas.

Question: What types of services?

Answer: As you know, the voice world is expanding and bulging with new services based on calling- and called-party identification. If a member of your family called to ask you to pick up something on the way home from work, but missed you at the office, that selected call could be transferred to another destination, be it a late meeting somewhere in the city—or even your car phone. Later calls will contain traveling identities that enable pri-

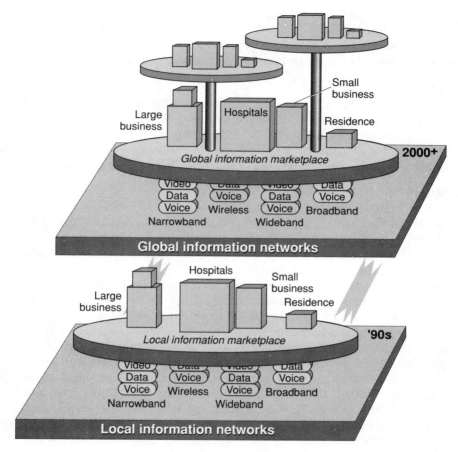

Fig. A-8. The telecommunication information millennium.

ority overrides, preempts, access to special survivable circuits for doctors during emergency situations, etc. The BOCs know and understand the voice network and its potential for new services such as voice mail, class, and voice-based intelligent services (800, 900, etc.); however, the real challenge is data.

Question: In what way?

Answer: RBOCs need to establish a new public data network that interconnects small business and residual terminals to large business, government, and education mainframe databases. It should enable doctors to search patient files, send prescriptions to the pharmacist, obtain information from poison control centers, and read X-rays at home. As you see, this crosses all of their market units: home, small and large business, state, federal, etc. Next we need to integrate data with voice so we better communicate. We need to be able to talk and look at the data while talking about it. During the conversion, we might want to change it and even see images derived from it. I believe we will want more databased image information in the future, but it must be current and available in "real time."

Question: How will they achieve this?

Answer: ISDN. This has long been a misunderstood term. I have many times said that I wish they had called it something else. ISDN is really an access vehicle to new services. It's the infrastructure for tomorrow. We have to come to realize that without this infrastructure, we do not have the appropriate base in place on which to layer our new information services platforms.

Question: Please explain.

Answer: ISDN is both an internetwork access vehicle to the new and different information networks, and the interconversational vehicle to enable dissimilar computers, terminals, and workstations to communicate with each other. Without ISDN, we will continue to do this in the private arena until we have a huge proliferation of different, loosely tied-together local area networks using specialized bridges, routers, and gateways.

This private internetworking trend is all right for the moment, for separate, specific information, but it becomes much more complex as the number of terminals increases and the volume of traffic expands in leaps and bounds. Then we need a more robust, versatile network. This is the challenge of the public information network. Frame relay and SMDS help interconnect LANs on a usage basis, and Internet has shown the need for being able to easily access addressable data terminals. But they have also shown the problems and pitfalls of not being as universal, as easy to use, and as secure and survivable as the fully switched voice network, with its complex and sophisticated network management traffic engineering and expandable networking hierarchical infrastructure.

We now need to initially deploy N-ISDN (or narrowband ISDN), which enables multiples of 64-Kbps channels to transport voice and data information throughout the network; however, to achieve success, we must not only deploy ISDN interfaces, but also establish the data network infrastructure to transport the data services. Then as we need to move more and more higher-speed data as images and full-motion video, we must take advantage of fully switched, addressable wideband W-ISDN capabilities and then B-ISDN (or broadband ISDN), as well as establish new network topologies that enable private local area networks to interface to each other through new public-network megabit and gigabit transport rings. Here new switching nodes are located closer to the user.

Hence, we need to first establish narrowband ISDN interfaces to our customers to move low- to medium-speed (64K/128-Kbps) data and images. Then, at the same time, we need to begin deploying fiber in these new topologies, using new automated mainframes as access switches near the user. We then need to establish new wideband/broadband switches to handle this high-volume, high-capacity traffic. Finally we need to establish new application service centers to provide a whole host of new data, image, and video services. Initially, these will be quite limited videotex-type gateways.

Question: Wow, this sounds like a whole new ball game.

Answer: Yes, we need to recognize that it's exactly that. ISDN could have also been called INIS (integrated networks' integrated services) or LNLS (layered networks' layered services). The key is to determine where one layer leaves off and another begins.

Question: Layered networks' layered services. Sounds interesting. How many layers? How will they look?

Answer: This is where the private and public internetworking takes place to enable interprocessing and interservices. There can be several ways of looking at it, but we see it as having at least five local layers.

The first layer is the CPE access layer—the PBXs, cluster controllers, cluster data switches, LANs or whatever. This interfaces to the second layer—the transport access layer, which has the transport access switch nodes that interface private-to-public facilities. These can be transport rings with the CPE on one side and the internal network switches on the other. Here the mainframe is automated and located closer to the customer. The third layer is the internal public switching network, where the call-handling functions are performed that are traditionally related to the class five. Data and video switching, translations, and customer database lookups are provided to achieve the more complex route-control and call-processing functions. In time, some of these functions will move to the transport access layer, and new functions will be added as higher layer functions are moved down. The next layer is where we will locate application service centers. Here are the gateways to videotex, 911 centers, special service database centers, special data-handling software systems, video database systems, etc. The final layer, the fifth layer, is on customer premises. This is the CPE application service center. Examples are internal data processing systems in hospitals or banks or whatever. Then there are layers six to ten for the IXCs and VANs' POPs, networks, and ASCs.

Question: This model looks realistic. What do we need to do to make it a reality?

Answer: The providers need to begin this type of dialogue with equipment suppliers to more clearly establish interfaces to these layers for new systems with Open Network Architecture (ONA) entrance/exit access points, using the ISDN OSI model and IEEE standard LAN/MAN interfaces. These application service centers are where third-party software can exist for information service providers (ISPs) and enhanced service providers (ESPs) delivering services that the appeals courts have not yet allowed the regulated telephone company to provide without extensive restrictions.

Question: Can the RBOCs make money on their new services?

Answer: If data networking and video communications hold the future promise for exciting arrays of new needed services that will be accepted by the general public, we need only to look at what a controversial and quite questionable offering—the 900/976 service—has experienced. In the early 1990s, after being initially exploited by the porn industry, it began to expand to a whole host of new offerings provided by 10,000 programs or so at prices from 50¢ per minute to $35 per month. An average call was 2.75 minutes at $1.25 per minute. Mid-1990 revenue estimates are in the high multibillions, as shifts are made to provide the following new services: weather forecast, sports results, tourist tips, product data, customer service, newspaper personal ads, stock tips, tax help, games, contests, sweepstakes, horoscopes, coons, polls, fund-raisers, pop-star trivia, auction news, waging advice, ski conditions, crossword hints, movie reviews, used car values, religious homilies, college admission requirements, diet support, pet help, missing children reports, insurance quotes, interest rates, legal assistance, the Easter Bunny, chit-chat, sleep expertise, horse-race results, job openings, and the capabilities for numerous folks yakking to each other across group-access bridge lines, which possesses a deliciously perfect acronym: GAB.

Question: From the public network perspective, what is happening? Where are we going? Where do we want to go? And how should we get there?

Answer: Let's address these questions in terms of IDN, ISDN, INIS, and LNLS, beginning with what's happening and where we are going. By the early 1990s, the RBOCs had slowed their purchases of digital base-satellite RSU (remote switch unit) systems for the rural and urban areas. Many suppliers did not economically offer ISDN solutions with enough financial incentives to cause a massive changeover to digital, especially in the rural environment for which IDN was originally designed. Nor did the suppliers offer very many data-handling capabilities with the push for ISDN interfaces. ISDN was only marketed on an "island basis" for mainly voice services (second line), not data. SS7 deployment was not tied to ISDN in order to take advantage of D-channel services.

The private world was left to their bridges and routers until the big push in 1992 to provide frame relay and SMDS with a few digital cross-connections. Broadband ATM was challenged by STM and hybrid offerings, thereby remaining in the laboratories and testbeds. Few high-bandwidth services were offered other than raw data transport for computer to computer or LAN to LAN. Fiber was deployed between central offices, as well as to feeder distribution points, but not to the home or business except on a selected basis. RBOCs used overlay techniques to collocate digital remote switch units with existing analog switches in order to provide ISDN to a small percentage of their customers. Few, if any, LNLS options were selected by the early 1990s, leaving an opportunity for CATV and private networks to separately provide parallel switched services to the RBOC customers.

Where do we want to go? How can we get there? Using overlay IDN with ISDN data-handling capabilities, some RBOCs can begin to establish public data networks. This technique might require number changes or second numbers for their customers, since few central offices are being fully changed to digital in the metro areas, as the RBOCs wait for broadband.

It's important to recognize that broadband will come in stages with the new nodal access switch (class 6) as the 1990s switch. It will be located on rings or collocated in central offices. It will initially provide wideband switched services up to 50 Mbps, and will later function as a front-end switch to the new fully broadband superswitches of the late 1990s. If priced correctly, videophone/picturephone and high-definition TV (HDTV) markets will begin to grow by the late 1990s through the turn of the century, requiring these new superswitches to provide these video services more ubiquitously. But this will not happen until the RBOCs participate in the new worlds of narrowband and wideband data and image movement over the 1990s. Large businesses will use full-motion video via compression techniques for selected customers over private networks, as many new private network providers move to full multimedia voice, data, image, and video offerings.

Rural locations can be upgraded with the wideband (class 6) system collocated at the COs, which can later be "homed" on a future regional superswitch. Customers will also obtain wideband/broadband services from their new premises switches (the information PBXs—class 7s). These can interface to the world through standard switched interfaces, as they offer internal narrowband/wideband/broadband switching and direct access to POPs, private application service centers, public SMDS offerings, and private global networks established by VANs.

By the mid-1990s, RBOCs will need to become more aggressive in addressing narrowband, wideband, and broadband planning and developments. They will need to work with their suppliers to purchase wideband switches, add ATM/STM capabilities to their existing switches, and define requirements for next-generation broadband switches, as they encourage interconnected private and public multimedia layered networks' layered services (LNLS) and establish more ubiquitous "information-networking" offerings in the late 1990s.

Question: What else needs to be done?

Answer: Providers need to design new maintenance, network management, and administrative systems to handle data and video traffic, and educate their service staffs to be able to support the new, more complex data/video traffic.

Question: This brings us to people.

Answer: Yes. They, of course, are key to all of these endeavors. We must re-train voice telecommunications people to understand the world of data, LANs, and video services in order to adequately market the new services to the more sophisticated customers, and enable public and private networks to be up and functioning.

Question: What will the cost be? Will RBOCs have the funds to build this infrastructure?

Answer: This is where the balancing of today's and tomorrow's needs must take place. We still need to satisfy today's stockholders. But to keep them, we must position ourselves to be in tomorrow's marketplace. It will take a lot of money to upgrade the voice network into an information network. We might need to do some creative financial planning with higher return on longer-term investments. Clearly, we must continue to meet the needs of today's stockholders. On the other hand, if LECs don't construct the new public infrastructure and service platform, the U.S. will not be a leader in the field tomorrow. We might not even be in the game.

Question: What changes do you see in the governmental/regulatory community that we need in order to play the new information game?

Answer: The United States needs to realize that it cannot be a successful player in the information age if its citizens cannot transport and process lots and lots of information. In the five years since divestiture, there have been very few incentives to bring this all together in order for it to happen. The longer that this continues, the further behind we become, and our industrial base becomes less competitive in the global marketplace. We could be building communications facilities that would alleviate our transportation problems.

We need to create new cities, especially in the large, expansive areas of the midwest and west. They can be tied together by the supergigabit facilities so that the quality of life is returned to our society as congestion, water, and waste problems become more manageable. It all remains to be seen.

The challenges of the 1990s are indeed formidable, but they appear to be essential in the transition from LECs being telephone companies, providing only voice services, to becoming information companies in tune with the marketplace. They will have to be able to provide an ever-increasing complex, array of voice, data, and video services, internetworking with the private communities. This indeed is their challenge of the 1990s. (See Fig. A-9.)

A final footnote

As noted earlier in the acknowledgements, this analysis completes the series. These new telecommunication information networks, products, and

Fig. A-9. Infoworld.

services offer an exciting array of new technologies that will have substantial impact on society—if used correctly. They will provide an awesome new power that can be used for the betterment or the destruction of society. The choice is ours.

> *Indeed, never in the history of humanity*
> *have so many technical possibilities*
> *been available to provide*
> *so many market opportunities*
> *as we enter a new millennium,*
> *the information millennium.*
> *Hopefully,*
> *we'll better manage our technologies*
> *to help us develop a better world,*
> *a safer world,*
> *a closer world,*

to each other and our God.
For a society
without morals has no values;
without values, there's no discipline;
without discipline, there's no order;
without order, there can be no society.
With this, I leave you, with a wistful toast
to everything that might have been,
to everything that was,
to everything that's yet to be.

Adieu

B

Technologies, standards, and services

"Terminology is 75% of any endeavor."

Horace Keith Heldman

Let's now summarize the technical aspects of achieving narrowband, wideband, and broadband transport. The networks are now being changed from analog networks that were based on representing voice over a 4000-cycle range of frequencies to a digital network in which the voice conversation is sampled 8000 times a second to determine the shape and amplitude level of the voice wave. This is called *quantizing* the analog voice and representing it in a stream of numbers. There are 256 numbers in the binary digital form of 1s and 0s representing the different shape and amplitude characteristics of the voice conversation. This is called *digitizing*. It takes an 8-bit code to indicate the position of the wave indicating the number from 0 to 256. Since this information is sent 8000 times a second, this requires $8 \times 8000 = 64,000$ bits per second (bps). This is called a *DS0*. When 24 conversations are sent together in a combined mode, we achieve 24 channels $\times$ 64,000 bits per channel = 1.536 million bits per second (Mbps). This payload, with an additional 8000 bits per second to achieve proper framing, (1 bit per frame 8000 times per second) = 1.544 Mbps. This is called a *DS1* or *T1*. Another way of looking at it is to say 8 bits per voice conversation $\times$ 24 conversations = 192 bits + 1 framing bit = 193 bits per frame $\times$ 8000 frames = 1.544 Mbps—8000 frame synchronization bits = 1.536 Mbps payload.

These frames are stacked together to form groups and supergroups of combined voice conversations having their own reference designations, such that DS1c = 2 DS1s = 3.152 Mbps. Also, DS2 = T2 = 96 voice conversations, or 4 T1s = 6.312 Mbps. Similarly, DS3 = T3 = 672 voice conversations = 28 T1s = 44.736 Mbps. T1s are multiplexed together to achieve T3s via M13 multiplexers. For 56-Kbps data traffic, 24 channels = 1.344 Mbps. This can be packaged in a T1. For primary-rate ISDN, called P-ISDN or W-ISDN, there are 23 B + D channels, where B = 64K and D = 64K and the total equals 1.544 Mbps. The D channel allows separation of voice and data by channel content identification. In narrowband ISDN (N-ISDN) = 2B + D. Here the B channel = 64K and the D channel = 16K for a total of 144 Kbps, which is encoded into a 160-Kbps bit stream to the central office. As noted earlier, voice is quantized and digitized into 64-Kbps bit streams and fits nicely into an ISDN channel. Note that adaptive PCM (ADPCM) uses a compression algorithm using voice in a 32-Kbps bit stream.

Synchronous Optical Network Transport (SONET) will package DS0, DS1, DS1c, and DS3 asynchronous bundles into a synchronous payload envelope (SPE) using a Synchronous Transport Signal protocol (STS-N), which has an optical carrier equivalent (OC-N). Europe uses a Synchronous Digital Network Hierarchy (SDH) represented by STM-N or CEPT-N levels of information transport, where the North American and European transport hierarchies for broadband information come together at 155 Mbps as well as 620 Mbps. These are considered the major customer interface rates for broadband ISDN (B-ISDN). Hence, there is a hierarchy of clusters for broadband SONET movement, such that STS-1 = OC-1 = 51.84 Mbps = 28 DS1 = 1 DS3. Alternatively, STS-3 = OC-3 = 155.520 Mbps = 86 DS1 = 3 DS3 = STM-4 while STS-12 = OC-12 = 622.088 Mbps = 336 DS1 = 12 DS3s, etc., on to STS-256, which is approximately 13 Gbps, etc.

The synchronous optical network utilizes wideband-broadband switches to switch and multiplex DS0, DS1, DS1c, DS3, STS-1, and STS-3 together. This enables combining and removing channels of information all the way down to an individual DS0. A virtual tributary (VT) is a container for sub DS3 payloads transporting signals below 50 Mbps. There are four VTs covering the different subrates below DS3. VT1.5 = 1.5 Mbps, while VT2 = 2.048 Mbps (European), VT3 = DS1c = 3.15 Mbps, and VT6 = DS2 = 6.34 Mbps. Similarly, VT groups contain all of one type of VT (for example, four VT1.5s or three VT2s or two VT3s or one VT6). Seven VTGs = one STS-1. STSs contain multiple numbers of DS0s, DS1s, DS2s, etc. or a DS3, where multiple STSs are concatenated together to form higher and higher rates.

SONET transfers information from end to end but uses a four-level model: physical, section, line, and path, where section terminating equipment (STE) signaling overhead ensures that information is transported properly between repeaters. Line terminating equipment (LTE) overhead handles origination and termination of line signaling across the network, and path-terminating equipment (PTE) overhead handles the communication between ends as information is multiplexed and demultiplexed. Thus the payload of varying sizes is moved from end to end along with transport overhead signaling.

The synchronous path envelope (SPE) is in effect nine rows of information where each row contains 90 bytes of information. The first three bytes of each row, called columns, contain transport overhead information. The first three rows contain section overhead, and the remaining six rows contain line overhead in the first three columns. The remaining 87 columns of bytes (where a byte is 8 bits) contain the information payload and have their own unique path overhead as well. Hence, 1.728 Mbps transport overhead + 50.112 Mbps payload = 51.840 Mbps = STS-1. Alternatively, $9 \times 90 = 810$ bytes per SONET frame $\times$ 1 frame per 125 microseconds $\times$ 8 bits per byte = 51.86 Mbps = STS-1 or OC-1.

In conclusion

SONET allows the full transport of multiples of T1, T1c, T2, and T3, as well as the subrate European 2.06 Mbps. It carries this information in virtual tributary (VT) or DS3 payloads using overhead pointers to properly move the information across the network and enable the adding and dropping of T1s and the pinpointing of DS0s. Once SONET is up and running, frame relay SMDS, FDDI, and other offerings simply become access protocols to the SONET network. Information will be switched over asynchronous transfer mode (ATM) or synchronous transfer mode (STM) systems requiring a new hierarchy of switching systems such that customer premises and switching nodes closer to customer premises become part of the traditional hierarchy. Here the new class 6 ring switch becomes the access node on the network side, enabling access to various interexchange carriers (IXCs) points of presence, the ability to send the information around the ring and get off or go to a different class 5 level switch in the event of network failure, or simply get off at another class 6 and go directly to a customer. In addition, sub-6 multiplexing devices as remote modules can assist the distribution and reception of information quite near a cluster of customers, both residential or business. In the future, customer systems such as PBXs will become information switches enabling a class-level (class 7) access to traditional common carriers (CC) and to value-added networks (VANs). As voice conversations are digitized into streams of 1s and 0s and combined with data (which is already in 1s and 0s) and video information (where pixels are represented in 1s and 0s), we can then achieve fully interactive, fully switched narrowband, wideband, and broadband service offerings. We can have fully multimedia videophone, imagephone, dataphone, and telephone services. As time progresses, we can move from enhancing the copper plant to handle narrowband ISDN to then switching wideband information in the subchannels of SONET up to DS3 rates. Then we will have full fiber distribution systems enabling the full range of multiples of STS/OC-N bit streams as we move to a world of higher-resolution, higher-capacity video imaging and conversations via videophones operating at 50–155 Mbps in every home, in every location, interconnected to every nation.

A wise man once said, "Terminology is 75% of any endeavor." The computer and communications industries have developed their own particular acronyms and terms, which are presented in Table B-1.

Table B-1. Technical acronyms

A/D	Analog-to-digital conversion
ADM	Add drop multiplexer
ADM	Asynchronous data module
ADPCM	Adaptive differential pulse code modulation
ANI	Automatic number identification
ANSI	American National Standards Institute
ASCII	American standard code for information interchange
AUTOVON	Automatic voice network
B	Bearer (channel)
B channel	Bearer Channel
B8RZ	Bipolar violation every 8th pulse to zero
B8ZS	Bipolar with 8 zero substitution
BISYNC	Binary synchronous communication
BOC	Bell Operating Company
BRI	Basic rate interface
bps	Bits per second
byte	8-bit segments
CAD	Computer-aided design
CAP	Competitive access provider
CBEMA	Computer Business Equipment Manufacturers Association
CCC	Clear channel capability
CCIA	Computer and Communications Industry Association
CCIS	Common channel interoffice signaling
CCITT	Consultative Committee International Telegraph and Telephone
CCS	Hundred call seconds
CCS	Common channel signaling
CCS7	Common channel signaling 7
CCSA	Common control switching arrangement
CD	Collision detection
CEPT	Conference of European Postal & Telecommunications Administration
CI II	Computer inquiry II
CI III	Computer inquiry III
CLASS	Customized local area signaling services
CPE	Customer-premises equipment
CPE	Customer-provided equipment
CSMA/CD	Carrier-sense multiple access/collision detection
CTX	Centrex
CUG	Closed user group
D	D channel
D/A	Digital to analog

DACS	Digital access and cross-connect system
dB	Decibel
D channel	Delta channel
dc	Direct current
DCC	Data country code
DCE	Data communication equipment
DCE	Data circuit-terminating equipment
DCO	Digital central office
DCP	D-channel processor
DDD	Direct distance dialing
DEMUX	Demultiplexer
DEO	Digital end office
DES	Data encryption standard
DID	Direct inward dialing
DIG	Digital
DP	Dial pulse
DPBX	Digital private branch exchange
DPKT	D-channel packet switching
DS-0	Digital signal level 0 (64 Kbps)
DS-1	Digital signal level 1 (1.544 Mbps)
DS-1c	Digital signal level 1C (3.152 Mbps)
DS-2	Digital signal level 2 (6.312 Mbps)
DS-3	Digital signal level 3 (44.736 Mbps)
DSX-0	DS-0 cross connect frame
ECSA	Exchange Carrier Standards Association
EDP	Electronic data processing
FIPS	Federal information processing standard
FSK	Frequency-shift keying
FT1	Fractional T1
FX	Foreign exchange
G	Giga—Prefix meaning one billion
GaAs	Gallium arsenide (crystalline material)
Gb/s	Giga (billion) bits per second
GOPL	"Good old" private line
GOS	Grade of service
H0	Higher-speed bearer channel (384 Kbps)
H11	Higher-speed bearer channel (1.5436 Mbps)
HASP	Houston Automatic Spooling Program
HDLC	High-level data link control
HDTV	High-definition television
Hz	Hertz
IC	Incoming call identification
ICA	International Communications Association
IDCMA	Independent Data Communications Manufacturers Association
IDDD	International direct distance dialing
IDN	Integrated digital network
IEEE	Institute of Electrical and Electronics Engineers
IFIPS	International Federation of Information Processing Society
INWATS	Inward wide area telecommunications service
I/O	Input/output
IO	Integrated optics
IP	Internal protocol
IP	Intelligent peripheral
ISDN	Integrated services digital network

Table B-1. Continued.

ISSN	Integrated special services network
ISUP	ISDN user part
ITU	International Telecommunications Union
IVDT	Integrated voice/data terminal
IXC	Interexchange carrier
JCL	Job control language
k	Kilo = 1,000
Kb/s	Kilo (thousand) bits per second
LADT	Local area data transport
LAN	Local area network
LAP	Link access procedure
LAPB	Link access protocol balanced
LAPD	Link access protocol delta channel
LASS	Local area signaling service
LATA	Local access and transport area
LCN	Logical channel number
LD	Laser diode
LDN	Listed directory number
LDS	Long-distance service
LEC	Local exchange carrier
LSI	Large-scale integration
LU6.2	IBM data communications protocol
m	Milli
M	Mega
MAN	Metropolitan area network
MDF	Main distributing frame
MF	Multifrequency
MFJ	Modified final judgment
MIPS	Million instructions per second
MODEM	Modulate/demodulate
MOS	Metal-oxide semiconductor
MTBF	Mean time between failure
MTTR	Mean time to repair
MUX/DEMUX	Multiplexer/demultiplexer
NANP	North American numbering plan
NBS/ICST	National Bureau of Standards/Institute for Computer Sciences & Technology
NCTA	National Cable Television Association
NCTE	Network channel-terminating equipment
NE	Network management
NE	Network equipment
NPA	Numbering plan area
NT	Network termination
NT1	Network termination type 1 (layer 1)
NTS	Network termination type 2 (layer 1–3)
NTE	Network terminal equipment
NTIA	National Telecommunications and Information Administration
OAM&P	Operations, administration, maintenance, and provisioning
OC-3	Optical carrier-signal level 3
OC-M	Optical carrier-level
O-E	Opto-electric
OEIC	Opto-electronic integrated circuit

OSI	Open systems interconnection
OSP	Outside plant
OTC	Operating telephone companies
OUTWATS	Outward wide area telecommunication service
P	Priority
P(A)BX	Private (automatic) branch exchange
PAD	Packet assembler/disassembler
PAM	Pulse amplitude modulation
PANS	Pretty awesome new service
PAX	Private automatic exchange
PBX	Private branch exchange
PC	Personal computer
PCM	Pulse code modulation
PCS	Personal communication service
PDM	Pulse duration modulation
PDN	Public data network
PNDP	Private network dialing plan
POP	Point of presence
POT	Point of termination
POTS	Plain old telephone service
PPB	Permanent packet B channel
PPD	Permanent packet D channel
PPDSL	Point-to-point digital subscriber
PPSN	Public packet-switched network
PPSS	Public packet-switched services
PPT	Point-to-point
PREMIS	Premises information system
PRI	Primary rate interface
PROM	Programmable read-only memory
PS	Packet switch
PSD	Packet-switched data
PSDC	Public-switched digital capability
PSDS	Packet-switched digital service
PSN	Packet-switched network
PSPDN	Packet-switched public data network
PSS	Packet-switch system
PSTN	Public-switched telephone network
PSU	Packet-switched unit
PTT	Postal Telegraph and Telephone Public Network
PUC	Public utility commission
PVC	Private virtual circuit
PVN	Private virtual network
PWM	Pulse width modulation
Q and R	Quality and reliability
QA/QS	Quality assurance/quality surveillance
QAM	Quadrature amplitude modulation
QSAM	Quadrature sideband amplitude modulation
RC	Recent change
RFD	Request for development
RFI	Request for information
RFP	Request for proposal
RJE	Remote job entry
ROM	Read only memory
RSM	Remote switching module
RSU	Remote switching unit
R/T	Receiver/transmitter

Table B-1. Continued.

RT	Remote terminal
SCC	Switching control center
SCP	Service control point
SDLC	Synchronous data link control
SDM	Space division multiplexing
SDN	Software defined network
SDS	Switched digital service
SLC	Subscriber loop carrier
SMF	Single-mode fiber
SN	Subscriber number
SNA	Systems network architecture
SO	Service order
SONET	Synchronous optical network
SP	Signal processor
SPC	Stored program control
SPOOL	Simultaneous peripheral operation online
SQL	Structured query language
SS7	Signaling system 7
SSP	Service switching point
STM	Synchronous transfer mode
STP	Signal transfer point
STS-1	Synchronous transport signal level 1
STS	Shared tenant services
STS-M	M STS-1 signals
SXS	Step-by-step
SYNTRAN	Synchronous transmission
T1	A U.S. standards committee
T1	1.544-Mbps digital carrier facility (DS-1 format)
T1A1	T1 Standards Committee on Performance/Signal Processing
T1C	3.152 Mbps digital carrier facility (DS-1C format)
T1C1	T1 Standards Subcommittee on Carrier-to-CPE Interfaces
T1D1	T1 Standards Subcommittee on ISDN
T1E1	T1 Standards Committee on Power and Protection Interfaces
T1M1	T1 Standards Subcommittee on OAM&P
T1P1	T1 Standards Committee on Planning & Program Management
T1S1	T1 Standards Committee on Services & Architectures
T1X1	T1 Standards Committee on Digital Hierarchy and Synch.
T2	6.312-Mbps digital carrier facility (DS-2 format)
T3	45-Mbps digital carrier facility (DS-3 format)
TA	Technical advisor (USITA)
Tb/s	Terabits per second (10 to the 12th bits per second)
TCAM	Telecommunications access method
TCM	Time compression multiplexing
TCM	Traveling classmark
TCP/IP	Transmission control protocol/internet protocol
TDM	Time-division multiplexing
TDMA	Time-division multiple access
TE	Terminal equipment
TE1	Terminal equipment type 1 (ISDN)
TE2	Terminal equipment type 2 (Non-ISDN)
TEI	Terminal endpoint identification
Telco	Telephone central office
TEN	Trunk equipment number
TOD	Time of day
TR	Technical reference
TRBLS	Troubles

TRIF	Technical Requirements Industry Forum
TSI	Time slot interchange
TTY	Teletypewriter
UNP	Uniform numbering plan
USITA	United States Independent Telephone Association
VAN	Value-added network
VAR	Value-added reseller
VCS	Virtual circuit switch
VF	Voice frequency
VGPL	Voice-grade private line
VLSI	Very large scale integration
VM	Virtual memory
VNL	Via net loss
VTAM	Virtual telecommunications access method
WATS	Wide area telephone service
WDM	Wavelength division multiplexing
XB	Crossbar (X-bar) switching system
ZCA	Zero called address

Table B-2.
Open Network Architecture (ONA) services

Basic Serving Arrangements (BSAs)

Analog Private Line—D.C. Channel Service
Analog Private Line—Low Speed Data Service
Analog Private Line—Voice Grade Service
Analog Private Line—Audio Service
Analog Private Line—Video Service
Digital Data Service
DS1 Service
DS3 Service
Packet Switching (X.25)
Packet Switching (X.75)
Simultaneous Voice and Data
Voice Grade-Line-Circuit Switched
- Digital Switched Service—Basic
- Feature Group A Service
- Flat Rate Line
- Foreign Central Office Service
- Foreign Exchange Service
- ISDN Basic Rate Access (2B+D)
- Measured Rate Lines
- Message Rate Lines
- PBX Trunks
- 2-Way Trunk with DID (Analog)
Voice Grade-Trunk-Circuit Switched
- 800 Service
- DID Service
- DID Switched Access Service
- Digital Switched Service-Advanced
- Feature Group B Service

Table B-2. Continued.

- Feature Group D Service
- ISDN Primary Rate Access (23B+D)

Basic Service Elements (BSEs)

ANI Forwarding
ANI Order Entry
Access Service Billing Information
Alternate Traffice Routing
Answer Supervision-Line Side
Automatic Loop Transfer (Automatic Protection Switching)
Automatic Number Identification (FGB)
Automatic Number Identification (FGD)
Backup/Redirection (Packet)
Bridging
CUG Incoming Access Barred (Packet)
CUG Outgoing Access Barred (Packet)
Call Transfer
Call Transfer of DID
Called Directory Number Delivery (DID)
Caller Identification-Number
Caller Identification-Bulk
Clear Channel Capability
Closed User Group
DID Trunk Queuing and Basic Announcement
Dial Call Waiting
Directed Call Pickup
Directed Call Pickup with Barge-In
Distinctive Alert
Fast Select Acceptance (Packet)
Flow Control Parameters (Packet)
Hunting
Improved Transmission Performance
Interface Group 6
Logical Channel (Packet)
Logical Channel Layout (Packet)
Make Busy
Market Expansion Line
Message Delivery Service
Multiple Network Addresses (Packet)
Multiple Port Hunt Group (Packet)
Multiplexing
Network Access Service (960)
Nonstandard Window Size (Packet)
Permanent Virtual Circuit (Packet)
Private Line Conditioning
Reverse Charge Acceptance (Packet)
Reverse Charge Option (Packet)
Secondary Channel
Selected Number Reverse Billing
Three-Way Calling
Traffic Data Report Service
Uniform Call Distribution

Complementary Network Services (CNSs)

Abbreviated Access/Activation (1 or 2 Digit)
Auto Call (Packet)
Call Forwarding-Busy Line

Call Forwarding-Busy Line (Expanded)
Call Forwarding-Busy Line (Programmable)
Call Forwarding-Busy Line/Don't Answer
Call Forwarding-Busy Line/Don't Answer (Expanded)
Call Forwarding-Don't Answer
Call Forwarding-Don't Answer (Expanded)
Call Forwarding-Don't Answer (Programmable)
Call Forwarding-Variable
Call Forwarding-Variable without Call Completion
Call Forwarding-Variable Remote Activation
Call Rejection
Call Trace
Call Waiting
Continuous Redial
Custom Ringing
Custom Ringing-Call Forwarding
Deluxe Call Waiting
Dual Telephone Coverage
Expanded Answer
Hot Line
Last Call Return
Message Waiting Indication-Audible
Message Waiting Indication-Visual
Priority Call
Selective Call Acceptance
Selective Call Forwarding
Speed Calling (8 Number)
Speed Calling (30 Number)
Warm Line

Table B-3. IEEE 802 standards

802	Standard for Local Area Networks for Computer Interconnection
802.1	Architecture & Overviews
802.1A	IEEE Local Area Networks: Computer Interconnection
802.1B	Local Area Systems: Network Management
802.2	Local Area Networks: Logical Link Control
802.3	Local Network for Computer Interconnection (CSMA/CD)
802.3A	Medium Attachment Unit and Baseband Medium Specification for Type 10BASE2
802.3B	Broadband Medium Attachment Unit and Medium Specification
802.3C	Local Area Networks: Repeater Unit
802.3D	Medium Attachment Unit and Baseband Medium Specification for Fiber Optic Inter-Repeater Unit
802.3E	Physical Signaling, Medium Attachment and Baseband Medium Specification, Type 1BASE5
802.3F	Physical Signaling, Medium Attachment and Baseband Medium Specification, Type 1BASE5 Multi-point Extension
802.4	Revision LAN: Token-Passing Bus Access Method
802.4A	Fiber Optic Token Bus
802.5	LAN: Token Ring Access Method and Physical Layer Specifications Alignment with ISO DIS 8802/5
802.5A	LAN: Station Management Revision
802.5B	LAN: Telephone Twisted Pair Media
802.5C	LAN: Token Ring Reconfiguration
802.5D	LAN: Interconnected Token Ring LANs

Table B-3. Continued.

802.5E	LAN: Token Ring Station Management Entity Specifications
802.5F	LAN: 16 Mbits Token Ring Operations
802.5G	LAN: Conformance Testing of ANSI/IEEE 802.5
802.5H	LAN: Operation of LLC III on Token Rings
802.6	Metropolitan Area Networks
802.7	Recommended Practices for Broadband LAN

Appendix C

Playing the information game

As public networks providers, both LECs and IXCs, enter the data transport arena with Frame Relay as one of their key services, it is time to pause and reflect on many of the issues and concerns that have surfaced with this type of offering. To put everything in perspective, let's review the history of LAN transport, since one key focus of this service is specifically LAN-to-LAN interconnection.

The private communication managers' unending pursuit for data transport was originally ignored by the PBX industry, forcing these managers to pursue any and all options to transport data. First they established customer-premise networks called local area networks (LANs), which were based on internal, addressable buses and token-ring structures. Then they used various wide area network (WAN) interconnection options between LANs, such as leased lines, dial-up voice-grade switched lines using low- speed analog modems, or full point-to-point T1 and T3 trunks.

Then in the mid to late 1980s, these private communication managers tried to build specialized private networks to interconnect their firms' distributed branch offices through mesh networks using nodal-point switching platforms that specifically switched T1 and T3 traffic. They attempted to integrate voice and data traffic, but they lost voice conversations and delayed data streams. This caused them to shift to data-only transport, leaving voice separate on the public network.

Over time, private network managers continued to push for more economical transport to connect their growing population of distributed LANs. LECs attempted to meet this need by providing an economic incentive to obtain varying amounts of transport in bundles—such as fractional T1's 364 Kbps or 768 Kbps, or fractional T3's variable number of T1s. Suppliers provided digital cross connects for public offering that enabled more dynamic provisioning to eliminate lengthy path setup delays, leaving it to customers to select the type and amount of T1, T3, fT1, or fT3 transport. However, customers soon found that as transport prices dropped and bandwidth became a commodity, price crossover points of 40–60% usage of T1 or T3 encouraged them to purchase the full T-carrier capabilities or go to a pay-as-you-use type of service. Hence, the search continued for cheaper transport such as usage-rate-based services, and along came Frame Relay.

Frame Relay is a connection-oriented "fast-packet" technology using only the lower layers of transport protocols. This speeds up network transport by removing many of the security and reception acknowledgments, leaving end-to-end checking to CPE or higher-layer protocols and requiring retransmission of erroneously received information end-to-end. Customers are provided a range of 56-Kbps, 64-Kbps, . . . 1.5-Mbps offerings, where they obtain a guaranteed information transport rate at a selected level and in some cases are allowed to compete for additional capacity as needed. CCITT (under its I.122 standard for additional packet mode bearer services) and ANSI (within its 88–224's Frame Relay Bearer Service) define Frame Relay interfaces. CCITT uses ISDN D-channel for Frame Relay services for X.25-like switched virtual circuits (SVCs), and ANSI focuses on permanent virtual circuits (PVCs) for mainly LAN-to-LAN interfaces. Initially, the Frame Relay interfaces of ANSI intentionally do not support voice or data. (These capability limitations are being reconsidered at this point in time.) SVCs' capabilities are analogous to higher-layer X.25 protocol connections, requiring call setup and teardown.

PVCs, on the other hand, are more like dedicated private facilities that are always there, where call setup and tear-down is implemented via commands from the network management system. At a higher layer (the network layer), frames can be exchanged between switches in the form of a packet. They can encapsulate the frame as a packet and route the packets by means of switched virtual circuits (SVCs), and as a PVC. The actual path that the packet takes depends more on existing conditions such as congestion, rather than on storage tables. Some networks for Frame Relay-PVC are using relatively inexpensive switches, as these vehicles are used as cross points rather than actual switches. More sophisticated switches will be required for SVC capabilities. The interexchange carriers are interfacing with the local exchange carriers to offer Frame Relay services. For example, they are using the 56K, 64K, or 1.5-megabit rate to effectively provide cross-country PVC networks. Congestion is an increasing problem as more and more users compete for Frame Relay services.

With Frame Relay came the communication providers' major policy shift in transport pricing. This was the beginning of offering new shared-usage facilities. This new emphasis on shared usage encourages customers to abandon previous offerings (leased lines, point-to-point T carrier) and move to usage-based services such as Frame Relay. This particular offering has customer incentives to purchase one level of service with a given transport throughput guaranteed, and then compete for additional transport, as needed, at a reduced cost. Of course, for the initial customers for Frame Relay, there is sufficient bandwidth to meet their additional needs, but as more and more customers flock to this cheaper offering, as they will no doubt do, congestion will raise its ugly head. This has been forecast by numerous Frame Relay network planners, and this will bring with it a host of new problems.

On a selected basis, it is easy to provide a limited offering for a limited few and be successful, especially as a service is offered in a controlled, limited area, in a limited way, such as PVC. A few specialists can support this offering, realizing, of course, that customers are simply being encouraged to abandon one service in favor of another. But when this offering is to be available anywhere at any request, enabling transport to anywhere at anytime, this will require a more massive undertaking. It will require a considerably more complex SVC network with sophisticated technical training in order to fully support the service ubiquitously.

This is especially true for a provider's network organization, which leverages from its unique capabilities to standardize an offering and provide its massive size to a massive deployment. However, if once it is geared up to deploy an offering and that offering is no longer needed or substantially changes, it requires considerable expense and time to drop the offering and turn to another. Similarly, to stop and quickly turn around an aircraft carrier is considerably more complex than working with a small destroyer. So it is with technology. If services such as Frame Relay are provided anywhere at any time without sufficient network and support equipment and technical training for operational personnel, there are considerable problems in establishing and maintaining a successful offering. If a service is incomplete and needs to be quickly changed, replaced, or supplemented by another, this is not an appropriate offering for a large network provider because it can become a massive, expensive undertaking. So in reviewing Frame Relay in light of the history of where the industry has been and how it got here concerning the current method of deployment, support, and capabilities, it is important to assess the following concerns and issues:

As traffic increases, as usage begets more usage, which begets more usage, there is more and more competition for spare capacity. Hence success begets failure as customers complain of loss of information or inability to obtain needed transport capacity, for one of the characteristics of Frame Relay is to simply drop bits and bytes of information once congestion and blockage occurs. This leaves the LEC in the state of having to support a traffic-sensitive service, and depending on where and how customers are collected, traffic engineering will have to extend into and throughout the local loop. This is a problem that PUC regulators in many states severely cautioned when traffic-sensitive concentrators were initially deployed in the voice world.

An opposite problem is the other type of Frame Relay user who only requires the low end of its capabilities. The service is used to provide only 56 Kbps, whereas it operates over a full range of 1.5 Mbps. The local loop is conditioned for 1.5M, but only 56 Kbps is required. Unfortunately, this could very easily be resolved by using less-expensive technologies that easily facilitate addressable, fully switched two-way offerings. Some users will simply move to Frame Relay because of its faster connectivity protocol, which drops a lot of the security and transport reception assurance acknowledgments (compared to X.25), but these users will return to fully switched medium-speed data technologies once their protocols are improved or eliminated by newer technologies.

Deployment of Frame Relay to an as-needed, as-required public—in an anywhere, anytime approach—causes severe operational difficulties to a network provider covering a vast territory without an overall plan and commitment for establishing a ubiquitous offering. Initially, Frame Relay was deployed for as-requested, point-to-point traffic. Next came the desire for fully switched services. These required more ubiquitous availability to be achieved by overlaying a grid of switching nodes specifically designed for handling Frame Relay traffic throughout the LEC's regulated territory. This is no small task in either case (PVC or SVC), and it is very expensive for SVC, especially for a type of service that is so traffic-call-mix dependent and congestion prone.

It is also important to note that two major shifts are already taking place. Computer suppliers first noted privately, and later publicly, that the future of LANs, as they exist in the early 1990s, will be quite limited once the new PBXs are deployed with full ATM/STM distributed switching fabrics. As noted earlier, the initial reason for a LAN was because the voice-world PBX ignored the data users' needs, thereby driving them to addressable bus and ring physical architectural topologies. This failing is addressed by the next generation of PBXs, called IBXs, for the turn of the century. This is especially applicable as the demand for handling multimedia terminal transport increases. Integrated voice, data, text, image, and video traffic from business workstations and home communications systems will have a profound effect on future communication needs. These systems require plesiochronous, two-way, symmetrical, synchronous information exchange, thereby moving data transport into the arena of fully integrated multimedia information transport. Similarly, large mainframes and distributed, powerful processors need to obtain and communicate with greater and greater quantities of information at higher and higher speeds—much more than just 1.5 Mbps.

When the more continuous-bit-rate, delay-sensitive, larger-bandwidth offerings such as videophone and videoconferencing traffic are then added to data Frame Relay traffic, this will totally explode sensitive Frame Relay data movement, especially as the large mainframes and sophisticated workstations attempt to take advantage of the "spare capacity" to obtain less expensive transport. This causes more and more bits and bytes to be lost, thereby further irritating data and multimedia customers.

Adding some relatively unsophisticated, inexpensive ATM switching nodes here and there on an ad hoc basis to help provide some switched (SVC) offerings, augmenting point-to-point (PVC), achieves only a limited capability that will, in time, be difficult to support. To attempt to use a limited point-to-point offering or specialized switch offering using small systems with limited survival capabilities is unrealistic, especially if it is to handle the new, rapidly exploding growth of information services. It is impractical to expect such limited offerings to be the building blocks upon which numerous services will be established, requiring the network providers to ensure that the system does not go down, that congestion does not occur, and that there is no blockage and loss of information. A public network provider does a good job at a massive level, but it must have the right services with its massive offerings. What is appropriate for a private networking arena is not necessarily applicable for the public networking arena. There is a difference between the two types of network offerings, and it is important to recognize the distinction.

Throughout the world, there is a need to interface wideband services using a versatile, robust infrastructure with highly sophisticated network management mechanisms. Providers cannot encourage customers on leased-line facilities having a 99.998% availability to change over to fully switched operations that, at times, might or might not meet their service needs. One might ask, "What is being built? Specifically, how much can be established on Frame-Relay-based PVC/SVC offerings?" This especially becomes a concern for technology that is specifically designed for medium-speed LANs (which themselves are changing), for terminals that are changing, for applications that are changing, and for technologies that are greatly changing transport rates and enabling throughputs that are much higher than 1.5 Mbps.

There is a need for a fully supported, highly sophisticated wideband switched service—more than just Frame Relay. However, Frame Relay is a viable 1.5-Mbps wideband interface service. Some wish to extend it to 45 Mbps, and many believe that it is simply an interface service to a future ATM/SONET-based architectural network. In addition, it is interesting to consider using Frame Relay protocols in narrowband ISDN (N-ISDN) "D" channel services to take advantage of its faster CPE-network interface capabilities for moving "B" channel packet information at 64/128-Kbps data rates.

Primary-rate ISDN provides 23 "B" channels of 64,000 bits per second (DSOs) and one "D" channel operating at 64,000 bps. This enables the identification of the content of each individual "B" channel within the 1.5-Mbps data stream, and this will be a viable alternative to Frame Relay once wideband channel switching is available.

There are levels and layers of complications in attempting to interface separate networks based on frames, ATM 53-byte packets, STM circuits, and SONET DSOs. There is a need, in fact an essential need, to understand where one network leaves off and another takes over, as well as where one is just an interface to another. This becomes especially important as narrowband technologies with limited conditioning of outside plant will enable both circuit and packet transport rates of 64K and 128K. SONET virtual tributaries can nicely package T1s, T2s, and T3s, enabling access to individual DSOs (64K channels) together with higher OC-N broadband ISDN rates of 155/622 Mbps. STM switching for continuous-bit-rate traffic for a variable number of channels can work together with ATM switching for bursty, variable-bit-rate traffic. Hence, it is important to know where one technology leaves off and the other takes over, if we are doing more than simply replacing a leased-line/trunk point-to- point LAN interface.

Finally, it is time to ask the following: What is a viable data transport network vision? Where is the telecommunications industry going with services such as Frame Relay? How should the industry use technologies such as ATM and STM? How will the families of new, expanding data services interrelate pricewise? With what cross-product impact in the marketplace? Where will the LECs and IXCs be in ten years? How shall regulators address these issues? Will private network managers simply abandon Frame Relay in the future, as they have with previous technologies, as they search for the cheapest transport until LANs are replaced, supplemented, or augmented by the next generation of fully distributed, full-bandwidth PBX technologies? And how will network providers successfully, ubiquitously deliver services in this type of changing competitive environment?

Epilogue
The information society

"Each age is an age that is passing, and cities, my friend, are transitory things. Each is born from the dust; each matures, grows old, then it fades and dies . . . A passing traveler looks at a mound of sand and broken stones and asks, 'What was here?' . . . and his answer is only an echo or a wind drifting sand."

Louis L'Amour

The old historian sat in front of his late-evening fire and reflected on the words he had just written for his upcoming speech at the International Historical Institute. He was visibly tired as he read the notes from his "think file." It was New Year's Eve. This time was always the loneliest time of the year without his wife, who had died some 20 years ago. His daughter would be having him over for dinner the next day after church. She had wanted him to spend the evening with them, but he preferred to continue the custom that Valerie and he had established of spending the eve of the new year in his library before an open fire. There they enjoyed looking back over the past and discussing the possibilities of all that was yet to come. Next year would be the end of the century. He would be 97 years old. He wished she was here to see it. Indeed, it had been an eventful century, one of considerable change, but more than that, perhaps the end of considerable turmoil—at least the end of the restructuring and repositioning, and hopefully, as some envision, the beginning of a period of stability. How had they arrived at this juncture in time? Where were they going? As his eyes grew more and more tired, he fumbled to look closer at his speech notes.

"As we draw closer to a new millennium, let's take a moment to look back over the past 3,000 years to determine what key events took place to cause significant shifts in our lifestyles—for their betterment—as we continued the neverending pursuit for a better quality of life for ourselves, our children, and our children's children. So here we are at 2999. What has happened to make life a little easier, a little better for all? Besides the many inventions in farming tools and manufacturing machines, we cannot ignore the alphabet's written word—obtained universally via the printing press. For without the initial Gütenberg Press, the western world could never have been exposed to new thinking and would not have had the opportunity to learn from the past—both past ideas and past mistakes. As we ventured forth from the days of learning by word of mouth or by reading manuscripts that were manually copied in the monasteries, it would have taken considerably more time, if it had been at all possible, to achieve the tremendous advances in scientific understanding that enabled the conquering of many diseases and the ability to travel farther and faster by land, sea, air, and space. Indeed, the printing press has been a boon for humanity, providing the catalyst for much to follow. Unfortunately, many inventions were also used in war and destruction, as seen by Nobel's dynamite, Ford's assembly-line tanks, and Einstein's nuclear-powered bomb. But, of course, great industrial and commercial uses have developed from these endeavors.

As we look back to the last century before our current millennium, we see that there was a mass movement from the rural farms to the urban cities during the industrial revolution of the twentieth century. Later, after a few centuries of false starts, with the advent of safe nuclear power from fission breeder reactors and the ability to harness direct sunlight via the moon, the world was able to take full advantage of the tremendous offerings of clean, inexpensive energy over the last half of the twenty-second century. Besides the expected uses in heating, lighting, and robotics, the initial far-reaching application was the ability to economically perform water desalinization. As unlimited fresh water became universally available in the twenty-third century, tremendous shifts in climate occurred, enabling the reclaiming of deserts by the reforestation programs of the twenty-fourth century. But of all these developments, the singular key achievement that enabled many new, exciting endeavors and covered the full range of computer-controlled applications was the successful deployment of global information telecommunications. This was quite different from the days of Alexander Graham Bell's universal telephone network.

As we look back over the last 1,000 years, especially the last 500 years of significant advancements, it is clear that these achievements in establishing the global information society could not have taken place if we had not obtained the highly sophisticated telecommunications networks by the end of the twenty-third century. Even in their earlier, somewhat primitive narrowband and wideband form, using the twisted-pair copper plant, they facilitated so much change that society made a marked shift in its mode of operation. The technology enabled many people to return to the somewhat empty rural environments, as remote regions of the world were suddenly able to have access to quantities of information that were usually only available for those living in the highly congested urban environments. These advances offset population growth concerns as people were able to spread out again from the tightly congested 1% of the land that they had been occupying for hundreds of years. Access to information—lots of it—enabled better control of valuable resources and waste management. Thus, as the fully fiberized broadband facilities matured with greater and greater throughput capabilities, astonishing developments occurred as distance constraints became meaningless.

After those early years of this millennium—now called the information millennium—we saw considerable change as we were able to better use our time, energy, and resources via tremendous communication improvements. This changed our way of life forever.

It enabled extensive penetration of computer technology into every aspect of society: inventory control, weather sensing, energy management, automating manufacturing plants, and robotics applications. By establishing universal, instantaneous electronic transmittal of all forms of information, technology eliminated many energy-consuming transport methods such as physically moving information (letters, documents, bills, etc.).

Technology helped people obtain a better home. Employees were able to work at home by forming the virtual corporation, in which distant employees networked together to design, produce, and market information-intensive products and services. New cities throughout the globe were formed. They were interconnected with the old cities via massive information highways, thereby achieving the phantom "global village" by the year 2350. The technology helped address, reduce, and resolve many of the more complex, challenging problems of the environment, poverty, and full employment."

Yes, the old man thought, substantial progress has been made, but what really happened over the past 1,000 years? Let's look a little deeper, he reflected, by breaking the millennium into two halves. He chuckled, noting that historians always like to summarize advancements in ages or periods. Well, the information age really began a millennium ago and proceeded in several phases. However, it took almost 300 years to really establish the communication structures needed to support the new era. Unfortunately, the preceding millennium had left a few unresolved major problems that needed to be simultaneously addressed as many new information network infrastructure advances were accomplished throughout this period. Historians still argue over cause and effect issues. If the communications networks had been established earlier, what would have been the resulting impact on society? But even if they had been, what would have been the outcome? Some believe you have to get ill, really ill, enough to scare you, before you go to the doctor and take your medicine. And sometimes it is too late, leaving the problems for those who come after you.

Each of the following areas could have been substantially affected by more efficient and accessible communications that better facilitated the more rapid exchange of information: education, banking, food production, medicine, traffic congestion control, law enforcement, the legal system, the building and construction industry, manufacturing, transportation, shipping, retail-wholesale trade, utilities, securities, financial exchanges, publishing, and entertainment.

Many needs for many years went unanswered: the lone protector on the street wishing to obtain more information from law-enforcement files on a particular suspect via description, eye print, thumb print, or voice print; the student in a rural farming community wishing to remotely sit in on a lecture at a major urban university; air traffic controllers needing better information on changing weather conditions; stranded freeway bus drivers wishing to know the traffic conditions on alternative highways. All of these people required the ability to access more and more timely data on their particular subject of interest. If more readily available and accessible, scientists could have made quantum jumps in the field of medicine, as researchers were better able to access, search, and share volumes of material that usually remained buried in the old untouchable archive files in the basement research morgue. Similarly, the design, manufacturing, and inventory control mechanisms involved in producing products could have been substantially improved to achieve more efficient, computerized production and global distribution.

So the old man reflected on what could have been, what would have been, what should have been, but wasn't because of the lack of foresight in the early years of the twenty-first century. All he could do was sigh and turn his thoughts to what indeed, unfortunately, did happen. Advancements were made, here and there in pockets, but considerably less than what was achievable at the time. Instead, without champions to foster the deployment, growth, and expansion of these expensive, new communication networks, substantial financial nearsightedness and territorial bickering between various market participants brought forth a delay that was to drag out for several centuries. Thus, the key window of opportunity was missed at the beginning of the third millennium. In its place, many crises developed as expanding population problems became more unmanageable and unresolved, as issues became more complex, more global.

Several disturbing events took place throughout the 2100–2600 time period. There was a 300-year drug war in which drug barons fought to establish and maintain their respective empires across the globe. During this time, governments grew exceedingly unstable and toppled, some at a rate of 20 or so per century, including governments of fully developed nations. By the end of the twenty-first century, terrorist activities and nuclear capabilities had become a disastrous combination. Before the universal ban, three major cities were partially destroyed by nuclear explosions until rational, concerned people united and stopped the warring parties. Unfortunately, chemical and biological warfare subsequently developed, and they were sometimes used against the populations themselves. These events, together with the outbreaks of deadly sexually transmitted diseases, including the rampant AIDS plague throughout the twenty-second

Fig. E-1. Growing populations.

century, caused considerable devastation of the population. It took an entire century to reduce the effects of these problems, and two more centuries to control them.

Crime was rampant until the twenty-fourth century, when the effects of the population shifts to the new cities became apparent. It took until the twenty-third century to reduce leadership resistance, especially in developed countries, to begin attempting more revolutionary, then evolutionary, changes. Things became quite desperate by 2100, with uncontrolled, exponential increases in crime, drugs, disease, negative environmental changes, traffic congestion, and food crises. Finally, stalled programs for reducing the size of the populated cities changed from being a possible alternative to a mandatory necessity. So after the initial 50 or so new cities were populated, efforts were made to fully establish the new information telecommunications infrastructure to provide a full spectrum of communication services to more than 10,000 rural locations, from which hundreds and hundreds of new cities sprang up all over the globe. After that, pressure mounted for rebuilding the old cities and interconnecting them in a huge telecommunications information grid.

With all these happenings, populations periodically grew and fell during the first 500 years or so, until around the year 2600. Then populations remained relatively the same. Once the many new cities were established, attention returned to really cleaning up the environment and using cheap energy to establish a 500-year program of sustained, stabilized growth. Unfortunately, there were a few glitches in the process, but overall the population was able to double and double again, enjoying a period of a relatively carefree, better quality of life—until we had the global nation wars of the past two centuries.

During this period, there was considerable positioning and repositioning as the world became segmented into ten viable nations. Some were called empires, others alliances, republics, or federations. Early in this process, we again suffered a major nuclear war, causing considerable devastation to the population. Several major areas of warring nations were annihilated over a short period. Finally, sanity again prevailed, shifting the competing forces to focus on the marketplace in fierce economic wars, where nations' economies were pitted against each other. Fortunately, we have now, hopefully, reached a period of relative stability.

So as we look to the future—the next millennium, the fourth millennium—we look to the new frontier—space. We had thought this to be the opportunity of the third millennium, but in actuality its challenge was to better occupy our own planet and stop the devastating abuses that technology offered when ignited with greed. By establishing the new information telecommunications networks, enabling layers and layers of more and more sophisticated computer services, we have indeed reclaimed the environment and greatly reduced crime, as we provided new cities to assist in upgrading the quality of life of the masses. Through this, technology helped conquer many diseases, both the old and the new biological diseases caused by new technology. Unfortunately, the rise in sexually transmitted diseases continued to play havoc with the public. So after the initial decline was offset by later advances, the majority of the populations returned to a more morally based, fully dispersed, less congested, globally interconnected mode of operation, enabling substantial growth in their quality of life. During the millennium, some space expeditions enabled the establishment of mining centers and sunlight power conversion plants on the moon and exploration centers on Mars, but now as the population approaches 25 billion and life is reasonably tranquil, eyes naturally turn to the next frontier, perhaps the last frontier—space.

The old man's thoughts were interrupted by a knock on the door. He arose and greeted an old friend, who made it a habit to visit him at this time to greet the new year, now that his wife had died and he also was alone. As they sat by the fire, their conversation returned to reviewing the accomplishments of past years and centuries. As usual, the two old historians ended up discussing the missing information telecommunications infrastructure in the twenty-first and twenty-second centuries. Why? Why did it not develop as fast as it could have? The discussion turned to government restrictions, incorrect management incentives, political concerns, personal motives, and the lack of a proper vision and the associated plans to achieve the vision. Many at that time suffered from economic nearsightedness or blindness rather than farsightedness. What could have happened if someone had taken the time to write down, at the end of the twentieth century, an exciting and feasible vision and plan for a global information society in the new telecommunications information millennium?

With that, the old man awoke. It was not the year 2999, but 1999. He had fallen asleep. Perhaps, he thought as he arose to go to bed, someone will take the time to write such a vision and plan. But, he reflected, who will read it? Who will help achieve it? But that is another story.

> *"Time present and time past*
> *Are both perhaps present in time future,*
> *And time future contained in time past.*
>
> **T.S. Elliot**

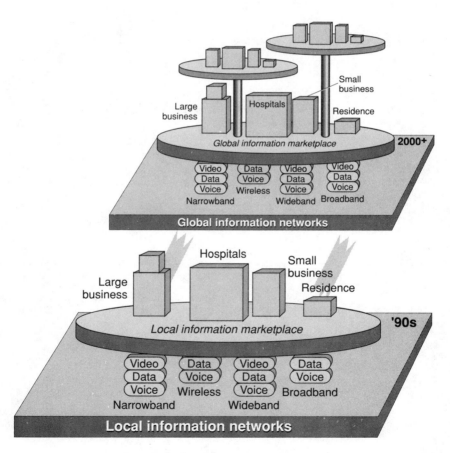

Fig. E-2. The information society.

Index

About the author

Robert K. Heldman is Director of Technical Strategies for U S WEST, where he has identified a new family of voice, data, and video information networks, products, and services for the 1990s. Previously he was Director of Technology Assessment for Northwestern Bell, where in his role of corporate chief scientist in the Strategic Planning Group, he helped establish postdivestiture positioning strategies for offering ISDN products and services in both the private and public marketplace.

Before joining Northwestern Bell, he was Manager of New Telecommunication Product Planning and Development for ITT's Advanced Technology Center. He was also responsible for defining the process of conceiving, defining, designing, producing, and supporting products for the world telecommunications and electronics industry. At the same time, he chaired an ITT multiunit system steering committee of conceptual planners who identified future digital integrated service networks' telecommunications products. The committee also specified technical areas in programming, switching, and technology that required further ITT research and development programs.

Prior to joining ITT, Mr. Heldman was both a Senior Conceptual Planner and Design Manager for GTE. There, along with several associates, he defined GTE's digital communication network for handling voice and data communications in the 1980s. He participated in writing the prospectus for GTE's digital switching product line. To support, modify, and refine his technical proposals, he performed extensive market research with leaders of 17 major industries in order to identify their future feature requirements. He subsequently spent two years with GTE's Northwest telephone company to better understand their operating problems in terms of future centralized maintenance and administration requirements.

On the development side at the Laboratory, he was Senior Manager of several large complex software and hardware developments, during which he supervised the technical design of switching system's call-processing translation programs, database management systems, switch markers, and overall system performance.

Before joining GTE, Mr. Heldman was with North American Rockwell (Autonetics), where after participating in the development of products using the first integrated circuit designs, he was responsible for a major portion of the integration of Minuteman Communication and Weapon Systems with Boeing, Sylvania, RCA, Motorola, and IBM.

After leaving Notre Dame University with a B.S. in electrical engineering, he subsequently received a master's degree in electrical engineering and a master's degree in engineering administration. He has authored 6 books and 20 major papers on the future telecommunications industry, as well as internal key strategic planning documents for GTE, ITT, and U S WEST top management. He also has published a series for *Telephone Engineering and Management* magazine on "Top-Down Planning." He is a guest lecturer at George Washington University's Continuing Education Division on the planning process for telecommunications providers, suppliers, and users. He has also been a lecturer for a course at both George Washington University and the University of Wisconsin on ISDN's integrated networks' integrated services. He has several major telecommunication patents, a professional engineering license, and he is a member of IEEE, the Executive Planning Institute, and the Planning Institute for the Future.

Other Bestsellers of Related Interest

Wireless Networked Communications
—*Bud Bates*
This book is a single source of information for telecommunications and LAN managers, network planners, and others who want a clear explanation of today's wireless communications systems and how to implement them.
ISBN 0-07-004674-3 $50.00 Hard

McGraw-Hill LAN Communications Handbook
—*Fred Simonds*
Starting with the basic concepts and progressing to specifics, this handbook helps readers understand what a LAN is and does. It discusses every aspect of LAN hardware and software to help users decide if they need a LAN and, if so, how to go about purchasing, installing, and running one.
ISBN 0-07-057442-1 $65.00 Hard

V Series Recommendations—2/E
—*Uyless Black*
Data communications professionals and anyone else interested in transmitting data over telephone networks must understand the V Series Recommendations-specifications for defining the transmission of data between computers via the telephone network-to which virtually all of today's modems and multiplexers conform.
ISBN 0-07-005592-0 $45.00 Hard

Future Telecommunications
—*Robert Heldman*
This book takes a realistic look at the applications and products in America's telecommunications future. For the first time, leading experts and business executives describe the full range of information applications that exist now, how they will change, and what narrowband, wideband, and broadband services and products will be required.
ISBN 0-07-028039-8 $29.95 Hard

Global Telecommunications
—Robert Heldman
This book gives a keen insight into the integration of world-wide private and public communications networks.
ISBN 0-07-028030-4 $40.00i Hard

Information Telecommunications
—Robert Heldman
This book gives a state-of-the-art analysis for everyone who must match up internetworking technologies and applications.
ISBN 0-07-028040-1 $45.00i Hard

Telecommunication Transmission Systems
—Robert Winch
This book provides the reader with a solid grasp of the technical details of today's digital transmission equipment, as well as valuable insights into future technological trends.
ISBN 0-07-070964-5 $70.00i Hard

Mobile Communication Satellites
—Tom Logsdon
This book fully explores the present status and future potential of the mobile communication satellite constellations currently being deployed in geosynchronous, low-altitude, and medium-altitude orbits.
ISBN 0-07-038476-2 $55.00i Hard

ISDN: Concepts, Facilities, and Services—2/E
—Gary C. Kessler
Aimed primarily at telecommunications and data communications specialists designing and implementing ISDN networks, this new edition also features many revised examples, more than 100 clear illustrations, and updated list of acronyms and abbreviations, and a complete glossary for easy reference.
ISBN 0-07-034247-4 $50.00 Hard

IBM Dictionary of Computing
—IBM
This book defines 18,000 terms and abbreviations, it's a one of a kind desk reference covering virtually all information processing, personal computing, telecommunications, office systems, and IBM-specific terms. No other dictionary of computing terms even comes close to the breadth of this one, nor is there anything available from such an authoritative source.
ISBN 0-07-031489-6 $24.95 Paper
ISBN 0-07-031488-8 $39.50 Hard

Open Computing's Guide to the Best Free Unix Utilities
—Keogh/Lapid
Includes text processing, games, communications, printing, spreadsheets,
development utilities, graphics utilities, electronic mail, and programming
languages. Key aspects of each program and utility are discussed along
with installation and operating instructions, as well as steps to modify the
utility to suit the reader's use.
ISBN 0-07-882046-4 $34.95 Paper/CD-ROM package

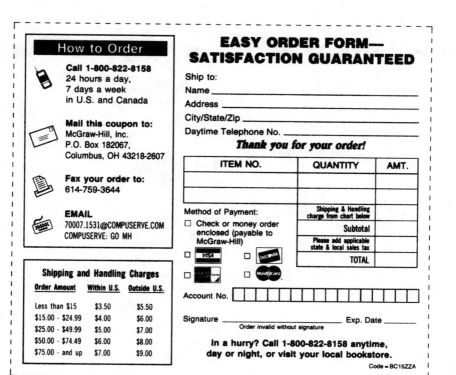